D0394025
NO LONGER PROPERTY OF
ANYTHINK LIBRARIES/
RANGEVIEW LIBRARY DISTRICT

Dreams of Earth and Sky

Dreams of Earth and Sky

Freeman Dyson

NEW YORK REVIEW BOOKS

New York

THIS IS A NEW YORK REVIEW BOOK

PUBLISHED BY THE NEW YORK REVIEW OF BOOKS

Published by The New York Review of Books, 435 Hudson Street, Suite 300,
New York NY 10014
www.nyrb.com

Library of Congress Cataloging-in-Publication Data

Dyson, Freeman J.
Dreams of earth and sky / by Freeman Dyson.
pages cm
ISBN 978-1-59017-854-6 (alk. paper)
1. Serendipity in science. 2. Discoveries in science. I. Title.
Q172.5.S47D97 2015
500—dc23
2014038482

ISBN 978-1-59017-854-6

Available as an electronic book; 978-1-59017-855-3

Printed in the United States of America on acid-free paper

1 3 5 7 9 10 8 6 4 2

Contents

Introduction

GREATEST BLUNDERS BY BOOK REVIEWERS

I AM GRATEFUL to *The New York Review of Books* for publishing this collection of my reviews from the years 2006–2014. It is a sequel to *The Scientist as Rebel*, which covered the years 1996–2006. The reviews in each volume are arranged in roughly chronological order. I put at the beginning of this one "Our Biotech Future," which is an essay and not a review. It was extracted from a lecture given at Boston University in 2005 with the title "Heretical Thoughts About Science and Society." I put at the end "The Case for Blunders," which happens to be my favorite.

Daniel Kahneman suggested the title for this introduction. That was his friendly response to "The Case for Blunders." In that review I gave him the wrong name, quoting one of his remarks and attributing it to David Kahneman. Somehow the "David" slipped unnoticed through three proofreadings. Kahneman's book *Thinking, Fast and Slow*, reviewed in chapter 16, explains how blunders of this kind happen. Each of us has two ways of thinking: the fast way for routine operations, and the slow way for situations requiring careful judgment. Authors are bad proofreaders because we tend to use the fast brain, impatient to get the job done quickly. The fast brain does not care about accuracy. The best proofreaders are professionals, paid by the hour and not by the page.

"David" is a small blunder. The big blunders in this book are not accidental but intentional. They are opinions that I hold in opposition to the prevailing wisdom. Since they are supported by the evidence that I can gather, I believe them to be true. Since they go against the majority view, I cheerfully admit that they may be wrong. *The New York Review of Books* gives me the opportunity to advocate views that are politically incorrect and provocative. I try to use this privilege sparingly, and I am grateful to readers who write letters correcting my mistakes.

Examples of big blunders in this collection are my sympathetic treatment of dubious characters such as Immanuel Velikovsky and Arthur Eddington (chapter 13) and William James and Sigmund Freud (chapter 16). Each of these characters built a universe of his own imagination outside the limits of conventional science, and each of them was shunned by the upholders of orthodox beliefs. I present them as heroes because I like to break down the barriers that separate science from other sources of human wisdom. Brilliant blunders break barriers and lead the way to a broader understanding of nature.

Another species of blunder that I treasure is concerned with politics rather than science. I am sympathetic to Wernher von Braun (chapter 3) and acclaim him as a hero, in spite of his membership in the SS and his complicity in the use of concentration camp victims to build his rockets. I oppose the idea, popular among my liberal friends, that war crimes should be prosecuted in perpetuity and never forgotten. History teaches us that after a war is fought to the bitter end, peace and reconciliation are more important than justice. Perpetuation of hatred and resentment is a chronic disease of human societies, and amnesty is the only cure.

My opposition to the prevailing wisdom concerning climate change and global warming is both a political and a scientific blunder. I do not claim to understand climate. I only claim that the experts who advise governments about climate also fail to understand it. There is a direct connection between my view of climate science and "The Case for Blunders." One of the blunders described in that review is the calculation of the age of the earth by William Thomson (Lord Kelvin) in 1862. Kelvin did a careful calculation based on his expert knowledge of physics and thermodynamics, ending with the result that the age should be about a hundred million years. We now know that the result was wrong by a factor of fifty. He got the wrong answer because he left out of his calculation some messy processes that he could not calculate, such as volcanic eruptions and lava flows.

In my view, the present-day calculations of global warming are similar to Kelvin's calculation of the age of the earth. The climate experts do careful and accurate calculations of computer models of the climate. The computer models are like Kelvin's picture of the earth, giving an accurate account of some processes and neglecting others. The computer models give an accurate account of the fluid dynamics of the atmosphere and ocean. They neglect some messy processes that they cannot calculate, such as the variable input of high-energy particles from the sun and the detailed behavior of clouds in the atmosphere. Darwin felt sure that Kelvin's calculation was wrong, because the evolution of life would require a time much longer than a hundred million years. I feel fairly sure that the modern calculations of global warming are wrong, because they do not give a good account of climate changes that occurred in the past. I am not claiming that the global warming calculations are wrong by a factor of fifty, but I would not be surprised if the predictions of future warming turned out to be wrong by a factor of five.

When science was in a creative phase, as it was in the nineteenth and twentieth centuries, there were various strongly held theories, some of which later turned out to be correct while others turned out to be blunders. Leading scientists argued passionately for their divergent views. Disputes among promoters of different ideas were essential to the process of understanding. In the end, nature spoke through observations that decided who was right and who was wrong. That is the way that healthy science moves forward. But it is not the way that climate science is moving now. Climate science has become politicized, so that one theory is officially declared correct and believers in other theories are silenced. That is why I question the official theory. I will accept it only after other theories have been publicly debated and rigorously tested. Debate and testing take a long time and cannot be hurried.

The review of John Gribbin's book *The Fellowship* (chapter 4) describes how, 350 years ago, the Royal Society of London laid a firm foundation for the integrity of science by adopting as its motto *Nullius in Verba*. This is a Latin phrase that educated people of that time could recognize as an abbreviated version of a well-known line of the poet Horace: "Sworn to follow the words of no master." In modern language, the Royal Society motto means "Nobody tells us how to think." When climate scientists cut short debate for political reasons, they are betraying their principles and forgetting their history.

I end this introduction with a review of the little book whose title I have borrowed. The book is *Dreams of Earth and Sky*, published in 1895 by the brilliant blunderer Konstantin Tsiolkovsky. His book is composed equally of science and science fiction, explaining to the general public the possibilities of space travel and space colonization. He was generally ignored for most of his life, living as a schoolteacher in the Russian provincial town of Kaluga, outside the academic and social hierarchy of the big cities. He lived long enough to become in his later years a Soviet hero, revered as the prophet and forerunner of the Soviet push into space.

When I recently went to see a Russian space launch at Baikonur, the historic center of the Soviet space program, I saw everywhere effigies of the Russian Holy Trinity of space: Konstantin Tsiolkovsky, the prophet who showed the way; Sergei Korolev, the chief designer of rockets; and Yuri Gagarin, the first human to fly in space. The Russian space culture is rooted in Tsiolkovsky's belief that we are on our way to the stars. The stars are our destiny. It may take us hundreds or millions of years to reach them, but we are on our way. Tsiolkovsky was not the only prophet of space travel. The Frenchman Jules Verne came before him and the German Hermann Oberth soon afterward. But Tsiolkovsky was the one who had the largest vision and the deepest understanding.

Tsiolkovsky's book tells us that, to be at home in the universe, we must solve two separate problems: one in engineering and one in biology. The engineering problem is the easy one. Tsiolkovsky worked out the mathematical theory of rockets and showed that rocketry would be a practical way for us to travel in space. He also explored solar sails as an alternative way to travel, slower but much less expensive. The biology problem is the hard one, to enable humans or other forms of life to be truly at home in the universe away from planets. The problem is to design living creatures that contain all the ecological resources of a planet within a small volume. In the science-fiction part of his book he describes his meeting with alien creatures. He calls them the natives and meets them strolling around on an asteroid.

The main subject of their conversation is whether small asteroids or planets are better as places to live. To the natives it is obvious that asteroids are better. They regard an atmosphere as an enormous hindrance, making it impossible to move without constant expenditure of energy to overcome atmospheric drag. The high gravity of a planet is also a big nuisance, causing additional waste of energy to overcome frictional forces when moving on the ground. To avoid being caught in a trap of frictional forces, they learned long ago to stay away from planets. For them, the small asteroids are the safest and most convenient places to visit in this corner of the universe. In the universe as a whole, small asteroids and not planets are the most likely places for life to evolve.

Since there is no sound in space, the natives communicate by sign language. Tsiolkovsky in real life was deaf, so he imagined himself mastering their sign language quickly and communicating with them better than he could communicate with humans on planet Earth. He was particularly interested in their anatomy and physiology. He observed that each native was both an animal and a plant, moving around with the brain and muscles of an animal, with life support provided by large green wings replacing lungs and stomach. The wings act like the leaves of a tree, using the energy of sunlight or starlight to drive all the chemical reactions that provide fuel for the brain and muscles. The wings have a skin without pores, unlike terrestrial leaves. Their skin is transparent and impermeable, not allowing any escape of air and water into space. To stay alive in space, everything inside the skin must be reused and recycled.

Tsiolkovsky calculates the wing area needed to sustain a closed ecology inside a native with a human-size brain and muscles at various distances from the sun. Only a small fraction of the incident solar energy is converted into chemical energy, the rest of it being used as heat to keep the native warm. He finds that the needed wing area is reasonable, equal to a few square meters for a native in the asteroid belt. If the wings grow thinner and wider to cover a much larger area, they can be used as solar sails. Evolution gives life the flexibility to adapt itself to various ecological niches in space, as it did on planet Earth. Given millions of years, life could have made the jump from planet to space, just as it made the jump from ocean to land. Tsiolkovsky saw the Earth as a tiny speck of dust in a vast universe. He saw our escape from imprisonment on this speck of dust to be desirable and in the end inevitable. He saw the freedom of space as our destiny. His vision is still alive in Russia and in some other places too.

The difference between the space cultures of the United States and Russia can be traced to the difference between the two pioneers, Robert Goddard and Konstantin Tsiolkovsky. The American pioneer Goddard was an engineer, and the American space culture is a culture of engineering. Tsiolkovsky was more concerned with biology than with engineering, and the Russian space culture is a culture of biology. The difference between engineering and biology causes a difference in the time scales of the two cultures. Americans tend to think of space programs with a time scale of years or decades. Russians, following Tsiolkovsky, tend to think with a time scale of centuries or millennia.

I borrowed Tsiolkovsky's title for this collection because hopeful dreams appear more frequently in the reviews than in the books. The wildness and wonder that pervade Tsiolkovsky's writing are rarely visible in recent books. Among the books reviewed here, only one, *The Age of Wonder* by Richard Holmes (chapter 9), captures the spirit of joyful dreaming that the modern world seems to have lost. Tsiolkovsky reminds us of the long-range dreams that our contemporary culture is lacking. Martin Luther King, only briefly mentioned in chapter 19, was a modern prophet who dared to dream. Nobody dreams now the way he did.

I

OUR BIOTECH FUTURE

IT HAS BECOME part of the accepted wisdom to say that the twentieth century was the century of physics and the twenty-first century will be the century of biology. Two facts about the coming century are agreed on by almost everyone. Biology is now bigger than physics, as measured by the size of budgets, by the size of the workforce, or by the output of major discoveries; and biology is likely to remain the biggest part of science through the twenty-first century. Biology is also more important than physics, as measured by its economic consequences, by its ethical implications, or by its effects on human welfare.

These facts raise an interesting question. Will the domestication of high technology, which we have seen marching from triumph to triumph with the advent of personal computers and GPS receivers and digital cameras, soon be extended from physical technology to biotechnology? I believe that the answer to this question is yes. Here I am bold enough to make a definite prediction. I predict that the domestication of biotechnology will dominate our lives during the next fifty years at least as much as the domestication of computers has dominated our lives during the previous fifty years.

I see a close analogy between John von Neumann's blinkered vision of computers as large centralized facilities and the public perception of genetic engineering today as an activity of large pharmaceutical and agribusiness corporations such as Monsanto. The public distrusts Monsanto because Monsanto likes to put genes for poisonous pesticides into food crops, just as we distrusted von Neumann because he liked to use his computer for designing hydrogen bombs secretly at midnight. It is likely that genetic engineering will remain unpopular and controversial so long as it remains a centralized activity in the hands of large corporations.

I see a bright future for the biotechnology industry when it follows the path of the computer industry, the path that von Neumann failed to foresee, becoming small and domesticated rather than big and centralized. The first step in this direction was already taken when genetically modified tropical fish with new and brilliant colors appeared in pet stores. For biotechnology to become domesticated, the next step is to become user-friendly. I recently spent a happy day at the Philadelphia Flower Show, where flower breeders from all over the world show off the results of their efforts. I have also visited the Reptile Show in San Diego, an equally impressive display of the work of another set of breeders. Philadelphia excels in orchids and roses; San Diego excels in lizards and snakes. The main problem for a grandparent visiting the reptile show with a grandchild is to get the grandchild out of the building without actually buying a snake.

Every orchid or rose or lizard or snake is the work of a dedicated and skilled breeder. There are thousands of people, amateurs and professionals, who devote their lives to this business. Now imagine what will happen when the tools of genetic engineering become accessible to these people. There will be do-it-yourself kits for gardeners who will use genetic engineering to breed new varieties of roses and orchids. Also kits for lovers of pigeons and parrots and lizards and snakes to breed new varieties of pets. Breeders of dogs and cats will have their kits too.

Domesticated biotechnology, once it gets into the hands of housewives and children, will give us an explosion of diversity of new living creatures, rather than the monoculture crops that the big corporations prefer. New lineages will proliferate to replace those that monoculture farming and deforestation have destroyed. Designing genomes will be a personal thing, a new art form as creative as painting or sculpture.

Few of the new creations will be masterpieces, but a great many will bring joy to their creators and variety to our fauna and flora. The final step in the domestication of biotechnology will be biotech games, designed like computer games for children down to kindergarten age but played with real eggs and seeds rather than with images on a screen. Playing such games, kids will acquire an intimate feeling for the organisms that they are growing. The winner could be the kid whose seed grows the prickliest cactus, or the kid whose egg hatches the cutest dinosaur. These games will be messy and possibly dangerous. Rules and regulations will be needed to make sure that our kids do not endanger themselves and others. The dangers of biotechnology are real and serious.

If domestication of biotechnology is the wave of the future, five important questions need to be answered. First, can it be stopped? Second, ought it to be stopped? Third, if stopping it is either impossible or undesirable, what are the appropriate limits that our society must impose on it? Fourth, how should the limits be decided? Fifth, how should the limits be enforced, nationally and internationally? I do not attempt to answer these questions here. I leave it to our children and grandchildren to supply the answers.

A New Biology for a New Century

Carl Woese is the world's greatest expert in the field of microbial taxonomy, the classification and understanding of microbes. He explored the ancestry of microbes by tracing the similarities and differences between their genomes. He discovered the large-scale structure of the tree of life, with all living creatures descended from three primordial branches. Before Woese, the tree of life had two main branches called prokaryotes and eukaryotes, the prokaryotes composed of cells without nuclei and the eukaryotes composed of cells with nuclei. All kinds of plants and animals, including humans, belonged to the eukaryote branch. The prokaryote branch contained only microbes. Woese discovered, by studying the anatomy of microbes in detail, that there are two fundamentally different kinds of prokaryotes, which he called bacteria and archea. So he constructed a new tree of life with three branches: bacteria, archea, and eukaryotes. Most of the well-known microbes are bacteria. The archea were at first supposed to be rare and confined to extreme environments such as hot springs, but they are now known to be abundant and widely distributed over the planet. Woese recently published two provocative and illuminating articles with the titles "A New Biology for a New Century" and (with Nigel Goldenfeld) "Biology's Next Revolution."*

Woese's main theme is the obsolescence of reductionist biology as it has been practiced for the last hundred years, with its assumption that biological processes can be understood by studying genes and molecules. What is needed instead is a new synthetic biology based

*See Carl Woese, "A New Biology for a New Century," *Microbiology and Molecular Biology Reviews*, June 2004; and Nigel Goldenfeld and Carl Woese, "Biology's Next Revolution," *Nature*, January 25, 2007. A slightly expanded version of the *Nature* article is available at http://arxiv.org/abs/q-bio/0702015v1.

on emergent patterns of organization. Aside from his main theme, he raises another important question: When did Darwinian evolution begin? By Darwinian evolution he means evolution as Darwin understood it, based on the competition for survival of noninterbreeding species. He presents evidence that Darwinian evolution does not go back to the beginning of life. When we compare genomes of ancient lineages of living creatures, we find evidence of numerous transfers of genetic information from one lineage to another. In early times, horizontal gene transfer, the sharing of genes between unrelated species, was prevalent. It becomes more prevalent the further back you go in time.

Whatever Woese writes, even in a speculative vein, needs to be taken seriously. In his "New Biology" article, he is postulating a golden age of pre-Darwinian life, when horizontal gene transfer was universal and separate species did not yet exist. Life was then a community of cells of various kinds, sharing their genetic information so that clever chemical tricks and catalytic processes invented by one creature could be inherited by all of them. Evolution was a communal affair, the whole community advancing in metabolic and reproductive efficiency as the genes of the most efficient cells were shared. Evolution could be rapid, as new chemical devices could be evolved simultaneously by cells of different kinds working in parallel and then reassembled in a single cell by horizontal gene transfer.

But then, one evil day, a cell resembling a primitive bacterium happened to find itself one jump ahead of its neighbors in efficiency. That cell, anticipating Bill Gates by three billion years, separated itself from the community and refused to share. Its offspring became the first species of bacteria—and the first species of any kind—reserving their intellectual property for their own private use. With their superior efficiency, the bacteria continued to prosper and to evolve separately, while the rest of the community continued its communal life.

Some millions of years later, another cell separated itself from the community and became the ancestor of the archea. Some time after that, a third cell separated itself and became the ancestor of the eukaryotes. And so it went on, until nothing was left of the community and all life was divided into species. The Darwinian interlude had begun.

The Darwinian interlude has lasted for two or three billion years. It probably slowed down the pace of evolution considerably. The basic biochemical machinery of life had evolved rapidly during the few hundreds of millions of years of the pre-Darwinian era, and changed very little in the next two billion years of microbial evolution. Darwinian evolution is slow because individual species, once established, evolve very little. With rare exceptions, Darwinian evolution requires established species to become extinct so that new species can replace them.

Now, after three billion years, the Darwinian interlude is over. It was an interlude between two periods of horizontal gene transfer. The epoch of Darwinian evolution based on competition between species ended about ten thousand years ago, when a single species, *Homo sapiens*, began to dominate and reorganize the biosphere. Since that time, cultural evolution has replaced biological evolution as the main driving force of change. Cultural evolution is not Darwinian. Cultures spread by horizontal transfer of ideas more than by genetic inheritance. Cultural evolution is running a thousand times faster than Darwinian evolution, taking us into a new era of cultural interdependence that we call globalization. And now, as *Homo sapiens* domesticates the new biotechnology, we are reviving the ancient pre-Darwinian practice of horizontal gene transfer, moving genes easily from microbes to plants and animals, blurring the boundaries between species. We are moving rapidly into the post-Darwinian era, when species other than our own will no longer exist, and the rules of open-source sharing will be extended from the exchange of software to the exchange of genes. Then the evolution of life will once again be communal, as it was in the good old days before separate species and intellectual property were invented.

I would like to borrow Woese's vision of the future of biology and extend it to the whole of science. Here is his metaphor for the future of science:

> Imagine a child playing in a woodland stream, poking a stick into an eddy in the flowing current, thereby disrupting it. But the eddy quickly reforms. The child disperses it again. Again it reforms, and the fascinating game goes on. There you have it! Organisms are resilient patterns in a turbulent flow—patterns in an energy flow.... It is becoming increasingly clear that to understand living systems in any deep sense, we must come to see them not materialistically, as machines, but as stable, complex, dynamic organization.

This picture of living creatures, as patterns of organization rather than collections of molecules, applies not only to bees and bacteria, butterflies and rain forests, but also to sand dunes and snowflakes, thunderstorms and hurricanes. The nonliving universe is as diverse and as dynamic as the living universe, and is also dominated by patterns of organization that are not yet understood. The reductionist physics and the reductionist molecular biology of the twentieth century will continue to be important in the twenty-first century, but they will not be dominant. The big problems—the evolution of the universe as a whole, the origin of life, the nature of human consciousness, and the evolution of the earth's climate—cannot be understood by reducing them to elementary particles and molecules. New ways of thinking and new ways of organizing large databases will be needed.

Green Technology

The domestication of biotechnology in everyday life may also be helpful in solving practical economic and environmental problems. Once a new generation of children has grown up, as familiar with biotech games as our grandchildren are now with computer games, biotechnology will no longer seem weird and alien. In the era of open-source biology, the magic of genes will be available to anyone with the skill and imagination to use it. The way will be open for biotechnology to move into the mainstream of economic development, to help us solve some of our urgent social problems and ameliorate the human condition all over the earth. Open-source biology could be a powerful tool, giving us access to cheap and abundant solar energy.

A plant is a creature that uses the energy of sunlight to convert water and carbon dioxide and other simple chemicals into roots and leaves and flowers. To live, it needs to collect sunlight. But it uses sunlight with low efficiency. The most efficient crop plants, such as sugarcane or maize, convert about 1 percent of the sunlight that falls onto them into chemical energy. Artificial solar collectors made of silicon can do much better. Silicon solar cells can convert sunlight into electrical energy with 15 percent efficiency, and electrical energy can be converted into chemical energy without much loss. We can imagine that in the future, when we have mastered the art of genetically engineering plants, we may breed new crop plants that have leaves made of silicon, converting sunlight into chemical energy with ten times the efficiency of natural plants. These artificial crop plants would reduce the area of land needed for biomass production by a factor of ten. They would allow solar energy to be used on a massive scale without taking up too much land. They would look like natural plants except that their leaves would be black, the color of silicon, instead of green, the color of chlorophyll. The question I am asking is, how long will it take us to grow plants with silicon leaves?

If the natural evolution of plants had been driven by the need for high efficiency of utilization of sunlight, then the leaves of all plants would have been black. Black leaves would absorb sunlight more efficiently than leaves of any other color. Obviously plant evolution was driven by other needs, and in particular by the need for protection against overheating. For a plant growing in a hot climate, it is advantageous to reflect as much as possible of the sunlight that is not used for growth. There is plenty of sunlight, and it is not important to use it with maximum efficiency. The plants have evolved with chlorophyll in their leaves to absorb the useful red and blue components of sunlight and to reflect the green. That is why it is reasonable for plants in tropical climates to be green. But this logic does not explain why plants in cold climates where sunlight is scarce are also green. We could imagine that in a place like Iceland, overheating would not be a problem, and plants with black leaves using sunlight more efficiently would have an evolutionary advantage. For some reason that we do not understand, natural plants with black leaves never appeared. Why not? Perhaps we shall not understand why nature did not travel this route until we have traveled it ourselves.

After we have explored this route to the end, when we have created new forests of black-leaved plants that can use sunlight ten times more efficiently than natural plants, we shall be confronted by a new set of environmental problems. Who shall be allowed to grow the black-leaved plants? Will black-leaved plants remain an artificially maintained cultivar, or will they invade and permanently change the natural ecology? What shall we do with the silicon trash that these plants leave behind them? Shall we be able to design a whole ecology of silicon-eating microbes and fungi and earthworms to keep the black-leaved plants in balance with the rest of nature and to recycle their silicon? The twenty-first century will bring us powerful new tools of genetic engineering with which to manipulate our farms and forests. With the new tools will come new questions and new responsibilities.

Rural poverty is one of the great evils of the modern world. The lack of jobs and economic opportunities in villages drives millions of people to migrate from villages into overcrowded cities. The continuing migration causes immense social and environmental problems in the major cities of poor countries. The effects of poverty are most visible in the cities, but the causes of poverty lie mostly in the villages. What the world needs is a technology that directly attacks the problem of rural poverty by creating wealth and jobs in the villages. A technology that creates industries and careers in villages would give the villagers a practical alternative to migration. It would give them a chance to survive and prosper without uprooting themselves.

The shifting balance of wealth and population between villages and cities is one of the main themes of human history over the last ten thousand years. The shift from villages to cities is strongly coupled with a shift from one kind of technology to another. I find it convenient to call the two kinds of technology green and gray. The adjective "green" has been appropriated and abused by various political movements, especially in Europe, so I need to explain clearly what I have in mind when I speak of green and gray. Green technology is based on biology, gray technology on physics and chemistry.

Roughly speaking, green technology is the technology that gave birth to village communities ten thousand years ago, starting from the domestication of plants and animals, the invention of agriculture, the breeding of goats and sheep and horses and cows and pigs, and the manufacture of textiles and cheese and wine. Gray technology is the technology that gave birth to cities and empires five thousand years later, starting from the forging of bronze and iron, the invention of wheeled vehicles and paved roads, the building of ships and war chariots, and the manufacture of swords and guns and bombs. Gray technology also produced the steel plows, tractors, reapers, and processing plants that made agriculture more productive and transferred much of the resulting wealth from village-based farmers to city-based corporations.

For the first five of the ten thousand years of human civilization, wealth and power belonged to villages with green technology, and for the second five thousand years wealth and power belonged to cities with gray technology. Beginning about five hundred years ago, gray technology became increasingly dominant, as we learned to build machines that used power from wind and water and steam and electricity. In the last hundred years, wealth and power were even more heavily concentrated in cities as gray technology raced ahead. As cities became richer, rural poverty deepened.

This sketch of the last ten thousand years of human history puts the problem of rural poverty into a new perspective. If rural poverty is a consequence of the unbalanced growth of gray technology, it is possible that a shift in the balance back from gray to green might cause rural poverty to disappear. That is my dream. During the last fifty years we have seen explosive progress in the scientific understanding of the basic processes of life, and in the last twenty years this new understanding has given rise to explosive growth of green technology. The new green technology allows us to breed new varieties of animals and plants as our ancestors did ten thousand years ago, but a hundred times faster. It now takes us a decade instead of a millennium to create new crop plants, such as the herbicide-resistant varieties of maize and soybean that allow weeds to be controlled without plowing and greatly reduce the erosion of topsoil by wind and rain. Guided by a precise understanding of genes and genomes instead of by trial and error, we can within a few years modify plants so as to give them improved yield, improved nutritive value, and improved resistance to pests and diseases.

Within a few more decades, as the continued exploring of genomes gives us better knowledge of the architecture of living creatures, we shall be able to design new species of microbes and plants according to our needs. The way will then be open for green technology to do more cheaply and more cleanly many of the things that gray technology can do, and also to do many things that gray technology has failed to do. Green technology could replace most of our existing chemical industries and a large part of our mining and manufacturing industries. Genetically engineered earthworms could extract common metals such as aluminum and titanium from clay, and genetically engineered seaweed could extract magnesium or gold from seawater. Green technology could also achieve more extensive recycling of waste products and worn-out machines, with great benefit to the environment. An economic system based on green technology could come much closer to the goal of sustainability, using sunlight instead of fossil fuels as the primary source of energy. New species of termite could be engineered to chew up derelict automobiles instead of houses, and new species of tree could be engineered to convert carbon dioxide and sunlight into liquid fuels instead of cellulose.

Before genetically modified termites and trees can be allowed to help solve our economic and environmental problems, great arguments will rage over the possible damage they may do. Many of the people who call themselves green are passionately opposed to green technology. But in the end, if the technology is developed carefully and deployed with sensitivity to human feelings, it is likely to be accepted by most of the people who will be affected by it, just as the equally unnatural and unfamiliar green technologies of milking cows and plowing soils and fermenting grapes were accepted by our ancestors long ago. I am not saying that the political acceptance of green technology will be quick or easy. I say only that green technology has enormous promise for preserving the balance of nature on this planet as well as for relieving human misery. Future generations of people raised from childhood with biotech toys and games will probably accept it more easily than we do. Nobody can predict how long it may take to try out the new technology in a thousand different ways and measure its costs and benefits.

What has this dream of a resurgent green technology to do with the problem of rural poverty? In the past, green technology has always been rural, based in farms and villages rather than in cities. In the future it will pervade cities as well as countryside, factories as well as forests. It will not be entirely rural. But it will still have a large rural component. After all, the cloning of Dolly the sheep occurred in a rural animal-breeding station in Scotland, not in an urban laboratory in Silicon Valley. Green technology will use land and sunlight as its primary sources of raw materials and energy. Land and sunlight cannot be concentrated in cities but are spread more or less evenly over the planet. When industries and technologies are based on land and sunlight, they will bring employment and wealth to rural populations.

In a country like India with a large rural population, bringing wealth to the villages means bringing jobs other than farming. Most of the villagers must cease to be subsistence farmers and become shopkeepers or schoolteachers or bankers or engineers or poets. In the end the villages must become gentrified, as they are today in England, with the old farmworkers' cottages converted into garages, and the few remaining farmers converted into highly skilled professionals. It is fortunate that sunlight is most abundant in tropical countries, where a large fraction of the world's people live and where rural poverty is most acute. Since sunlight is distributed more equitably than coal and oil, green technology can be a great equalizer, helping to narrow the gap between rich and poor countries.

My book *The Sun, the Genome, and the Internet* (1999) describes a vision of green technology enriching villages all over the world and halting the migration from villages to megacities. The three components of the vision are all essential: the sun to provide energy where it is needed, the genome to provide plants that can convert sunlight into chemical fuels cheaply and efficiently, the Internet to end the intellectual and economic isolation of rural populations. With all three components in place, every village in Africa could enjoy its fair share of the blessings of civilization. People who prefer to live in cities would still be free to move from villages to cities, but they would not be compelled to move by economic necessity.

Note added in 2014: This essay provoked a number of readers to write angry letters in response. Here is an example, published in The New York Review *of September 27, 2007, together with my reply:*

To the Editors:

Science is valuable and admirable for its ability to establish a certain kind of truth beyond a reasonable doubt, for its precise methodologies and its respect for evidence. And so it is disconcerting to see an eminent scientist such as Freeman Dyson using his own prestige and that of science as a pulpit from which to foretell the advent of yet another technological cure-all.

In his essay "Our Biotech Future," Mr. Dyson sees high technology as "marching from triumph to triumph with the advent of personal computers and GPS receivers and digital cameras," and he foretells the coming of a "domesticated" biotechnology that will become the plaything and art form of "housewives and children," that "will give us an explosion of diversity of new living creatures, rather than the monoculture crops that the big corporations prefer," and will solve "the problem of rural poverty."

This of course is only another item in a long wish list of techno-scientific panaceas that includes the "labor-saving" industrialization of virtually everything, eugenics (the ghost and possibility that haunts genetic engineering), chemistry (for "better living"), the "peaceful atom," the Green Revolution, television, the space program, and computers. All those have been boosted, by prophets like Mr. Dyson, as benefits essentially without costs, assets without debits, in spite of their drawdown of necessary material and cultural resources. Such prophecies are in fact only sales talk—and sales talk, moreover, by sellers under no pressure to guarantee their products.

Mr. Dyson has the candor to admit that biotechnological games for children may be dangerous: "The dangers of biotechnology are real and serious." And he lists a number of questions—serious ones, sure enough—that "need to be answered." But perhaps the most irresponsible thing in his essay is his willingness to shirk his own questions: "I do not attempt to answer these questions here. I leave it to our children and grandchildren to supply the answers." This is fully in keeping with our bequest to our children of huge accumulations of nuclear and chemical poisons. And isn't it rather shockingly unscientific? If there is anything at all to genetics, how can we assume that our children and grandchildren will be smart enough to answer questions that we are too dull or lazy to answer? And after our long experience of problems caused by industrial solutions, might not a little skepticism be in order? Might not, in fact, some actual cost accounting be in order?

As for rural poverty, Mr. Dyson's thinking is all too familiar to any rural American: "What the world needs is a technology that directly attacks the problem of rural poverty by creating wealth and jobs in the villages." This is called "bringing in industry," a practice dear to state politicians. To bring in industry, the state offers "economic incentives" (or "corporate welfare") and cheap labor to presumed benefactors, who often leave very soon for greater incentives and cheaper labor elsewhere.

Industrial technology, as brought-in industry and as applied by agribusiness, has been the cleverest means so far of siphoning the wealth of the countryside—not to the cities, as Mr. Dyson appears to think, for urban poverty is inextricably related to rural poverty, but to the corporations. Industries that are "brought in" convey the local wealth *out*; otherwise they would not come. And what makes it likely that "green technology" would be an exception? How can Mr. Dyson suppose that the rural poor will control the power of biotechnology so as to use it for their own advantage? Has he not heard of the patenting of varieties and genes? Has he not heard of the infamous lawsuit of Monsanto against the Canadian farmer Percy Schmeiser? I suppose that if, as Mr. Dyson predicts, biotechnology becomes available—cheaply, I guess—even to children, then it would be available to poor country people. But what would be the economic advantage of this? How, in short, would this *work* to relieve poverty? Mr. Dyson does not say.

His only example of a beneficent rural biotechnology is the cloning of Dolly the sheep. But he does not say how this feat has benefited sheep production, let alone the rural poor.

Wendell Berry
Port Royal, Kentucky

I replied:

My thanks to Wendell Berry for his illuminating comments. As usual, I learn more from critics than from flatterers. I value Berry's criticism especially because it comes from Kentucky, a state that I know only superficially from a visit to Center College in Danville, where I was a guest of the local chapter of Phi Beta Kappa students. In Danville I saw three things that agree with my vision of the future: a world-class performance of the Verdi Requiem by a local choir, a bookstore where the owners know and love what they are selling, and a roomful of bright students arguing about science and technology in the midst of a rural society.

I am aware that Danville is not all of Kentucky, and that large parts of Kentucky do not enjoy the blessings of gentrification. But I still see Danville as a good model for the future of rural society, when people are liberated from the burdens of subsistence farming. I am not foretelling any "technological cure-all." I am only saying that science will soon give us a new set of tools, which may bring wealth and freedom to the countryside when they become cheap and widely available. Whether we greet these new tools with enthusiasm or with abhorrence is a matter of taste. It would be unjust and unwise for those who dislike the new tools today to impose their tastes on our grandchildren tomorrow.

2

WRITING NATURE'S GREATEST BOOK

IVAR EKELAND HAS a Norwegian name and teaches at the University of British Columbia in Canada, but the style and spirit of his book *The Best of All Possible Worlds: Mathematics and Destiny* are unmistakably French.* The book is a rapid run through the history of the last four hundred years, seen with the eyes of a French mathematician. Mathematics appears as a unifying principle for history. Ekeland moves easily from mathematics to physics, biology, ethics, and philosophy. The central figure of his narrative is the French savant Pierre de Maupertuis (1698–1759), a man of many talents, who formulated the principle of least action in 1745 in a memoir with the title *The Laws of Motion and Rest Deduced from a Metaphysical Principle.* The principle of least action says that nature arranges all processes so as to minimize a quantity called action, which is a measure of the effort required to bring the processes to completion. The action of any mechanical motion is defined as the moving mass multiplied by the velocity and by the distance moved. Maupertuis was able to demonstrate mathematically that if a collection of objects moves in such

*University of Chicago Press, 2006.

a way as to make the total action as small as possible, then the movement obeys Newton's laws of motion. Thus the whole science of Newtonian mechanics follows from the principle of least action.

Maupertuis was dazzled by the beauty of his discovery. "How satisfying for the human spirit," he wrote, "to contemplate these laws, so beautiful and simple, which may be the only ones that the Creator and Ordainer of things has established in matter to sustain all phenomena of this visible world." He went on to identify action with evil, so that the principle of least action became a principle of maximum goodness. He concluded that God has ordered the universe so as to maximize goodness. The world that we live in is the best of all the possible worlds that God might have created. This simple principle unites science with history and morality. Mathematics is the key to the understanding of human destiny.

One of the contemporaries of Maupertuis was Voltaire, the great skeptic, who demolished Maupertuis's optimistic philosophy in a book with the title *The Story of Doctor Akakia and the Native of Saint-Malo*. *Akakia* is Greek for "absence of evil," and the native of Saint-Malo is Maupertuis. "The native of Saint-Malo," Voltaire writes, "had long fallen a prey to a chronic sickness, which some call philotimia [Greek for love of honors] and others philocratia [Greek for love of power]." Voltaire's book sold well and Maupertuis's day of glory ended. After Maupertuis died, Voltaire made him posthumously ridiculous by writing the novel *Candide*, in which Maupertuis appears as the optimistic philosopher Pangloss, wandering from one disaster to another but unshaken in his belief that "all is well that ends well in the best of all possible worlds."

Maupertuis was in fact no Pangloss. He spent only a small part of his time as an optimistic philosopher. He was also a brilliant scientist and a capable administrator. He became famous as a young man for leading an expedition to Lapland to measure the shape of the earth at high latitude. His measurements were accurate enough to prove that the earth is not a perfect sphere but an ellipsoid, flattened at the poles as Newton predicted as a consequence of its rotation. This confirmation of Newton's theory was historically important, since up to that time Newtonian physics was not widely known or accepted in France. Maupertuis also learned to travel on skis in Lapland, and brought home with him the first pair of skis that had ever been seen in France. For many years after the Lapland expedition, he was one of the most active members of the French Academy of Sciences. When King Frederick the Great of Prussia founded his own Academy of Sciences in Berlin, he invited Maupertuis to be the first president. Maupertuis spent the rest of his life in Berlin, successfully launching and running the Prussian Academy. Voltaire hated King Frederick, and Maupertuis's friendship with the king gave Voltaire another reason to hate and belittle Maupertuis.

Ekeland's sketch of history is divided into two parts: before Maupertuis and after Maupertuis. Before Maupertuis, the two chief characters are Galileo and René Descartes. Galileo started modern science by using the pendulum as a tool to make accurate measurements of time. Ancient Greek science was based on geometry, measuring space but not time. Archimedes understood statics but did not understand dynamics. Galileo with his pendulum and his falling weights made the decisive step from a static to a dynamic view of nature. He introduced time as a quantity accessible to mathematical analysis. He said, "Nature's great book is written in mathematical symbols." That remark by Galileo was the lever that moved the world into the modern era of scientific understanding.

After Galileo came Descartes, a great mathematician and a great philosopher but not yet a great scientist. Descartes took to heart Galileo's insight that mathematics is the language that nature speaks. He tried to deduce the laws of nature from the laws of mathematics by pure reason alone. He did not listen to another statement by Galileo, that nature answers questions that we ask by doing experiments. Descartes held experimental results in low esteem, thinking them less trustworthy than logic. His was a normative science, telling nature what it was supposed to do, and not an experimental science, investigating what nature was actually doing. In 1637 Descartes published his great work, *A Discourse on the Method of Rightly Conducting the Reason and of Seeking Truth in the Sciences.* He describes a scientific method that is broad enough to deal with moral as well as with physical problems. "I showed what the laws of nature were," he wrote,

> and without basing my arguments on any principle other than the infinite perfections of God, I tried to demonstrate all those laws about which we could have any doubt, and to show that they are such that, even if God created many worlds, there could not be any in which they failed to be observed.

Ekeland concludes that Descartes's method "has been used in science with tremendous success, and there is no reason why it should not be as useful in philosophy, or in trying to establish some principles by which to guide our collective and individual lives." Unfortunately, the Cartesian way of doing science with minimum recourse to experimentation led him into bad mistakes. From his philosophical principle that nature abhors a vacuum, he was led to deduce that the space around the planets is filled with enormous vortices, or whirling masses, and that the pressure of the vortices confines the planets to their orbits and pushes them on their way. This theory of planetary motions was generally accepted in France as a preferable alternative to Newton's theory of universal gravitation. Descartes also deduced that the rotating earth creates another enormous vortex that squeezes the earth into the shape of an American or rugby football. According to Descartes, the earth should be an ellipsoid elongated at the poles, instead of being flattened as predicted by Newton. Maupertuis's measurements in Lapland proved Newton right and Descartes wrong.

Ekeland's history continues after Maupertuis with a couple of great mathematicians—Joseph-Louis Lagrange and Henri Poincaré, who used the ideas of Maupertuis to build a grand edifice of classical dynamics. Poincaré, in the late nineteenth century, discovered chaos, a general property of dynamical systems that makes their behavior unpredictable over long times. He discovered that almost all complicated dynamical systems are chaotic. In particular, the orbital motions of planetary systems with more than two planets, and the fluid motions of atmospheres or oceans, are likely to be chaotic. The discovery of chaos opened a new chapter in the history of astronomy and meteorology, as well as in the history of mathematics.

After his discussion of Poincaré, Ekeland devotes chapters to biology and ethics, with backward glances to establish connections with Maupertuis. In biology, the guiding principle of evolution is the survival of the fittest. Darwin's notion of nature selecting a population with maximum fitness resembles Maupertuis's notion of God selecting a universe with maximum goodness. Darwin himself understood that fitness is not the same as goodness, but other evolutionary thinkers such as Herbert Spencer allowed the distinction between fitness and goodness to be blurred. Darwin rarely used the word "evolution," which Spencer introduced into biology. Darwin preferred to speak of "descent with variation," emphasizing the fact that variations are random and not usually progressive.

In ethics, the problem of optimization is even more tricky. Ekeland begins his discussion of ethics with Jean-Jacques Rousseau, the philosopher of the French Enlightenment, whose ideas prepared the way for the revolution of 1789. Rousseau believed that human beings were naturally virtuous and wise. They needed only to be set free from tyrannical governments, and then they would order their affairs harmoniously. A democratic government, responsive to the will of a free people, would make sure that everyone was treated fairly. Before the revolution could put these ideas to a practical test, some theoretical difficulties were raised by the Marquis de Condorcet, who for the first time used mathematics to model human behavior. The marquis discovered a logical inconsistency known as Condorcet's paradox, which demonstrates that an assembly ruling by majority vote may make decisions that are logically incompatible. For example, if three candidates A, B, C are running for a job to be filled by majority vote, it is possible that a majority prefers A to B, another majority prefers B to C, and a third majority prefers C to A. Then the result of the election will depend on the order in which the votes are taken. Another learned academician, the Chevalier de Borda, devised a system of preferential voting for election of members to the French Academy of Sciences. The Borda scheme avoided the Condorcet paradox, but led to another paradox that could be exploited by unscrupulous politicians to win elections. It turned out that no system of voting is free from mathematical paradoxes. And the revolution, when it came, brought a quarter-century of death and destruction instead of the peace and harmony that Rousseau had promised.

To sum up the lessons to be learned from history, Ekeland writes:

> We have now reached the end of our journey. It started in the world of the Renaissance, impregnated with Christian values....The laws of nature then are simply the rules God followed when creating the world, and the purpose of science is to recover them from observations. There is then also a deeper science, which is to seek the purpose God himself had in creating the world. This is what Maupertuis, in a glorious moment,

> thought he had achieved, thereby reconciling forever science and religion, both being the quest for God's will, in the physical world and in the moral one. Our journey ends in a world where God has receded, leaving humankind alone in a world not of its choosing.

While reading this account, I became more and more intrigued by the question of how a Norwegian working in Canada acquired a view of mathematics and of history that is so quintessentially French. The characters in his story are mostly French, and the dominant role of mathematics in their thinking is a hallmark of French culture. Nowhere else except in France do mathematicians command such respect. As soon as I consulted Google, I found the solution to the mystery. In spite of his Norwegian name, Ekeland is French. Born in Paris, educated at the historic École Normale Supérieure, a professor at the University of Paris–Dauphine, and subsequently president of the university, he is a charter member of the French academic establishment. His books were mostly written in French before being published in other languages. This book is a translation of a book with the same title published in French in 2000, revised and brought up-to-date for English-speaking readers. It gives us a vivid picture of human history and destiny as seen through the eyes of a senior academic trained in the French educational system.

There is at least one Frenchman who does not share Ekeland's view of the world. Pierre de Gennes is a brilliant French physicist who won a Nobel Prize in 1991 for understanding the behavior of squishy materials on the borderland between liquid and solid. He called the things that he studied "soft matter." After the Nobel Prize made him a French national hero, he was inundated with invitations to visit high schools and inspire the students to follow in his footsteps. He accepted the invitations and spent a year and a half as a traveling guru, explaining science to the kids. He enjoyed the contact with young people so much that he turned his talks into a book, *Fragile Objects: Soft Matter, Hard Science, and the Thrill of Discovery*. The book was translated into English and published by Springer in 1996. It describes in simple words how the science of soft matter explains the behavior of ordinary materials such as soap, glue, ink, rubber, and flesh and blood that children encounter in their everyday lives. De Gennes's talks were aimed at the average child, not at the talented few who might become professional scientists. His book is well pitched to give average readers a practical understanding of how science works.

At the end of his book, de Gennes adds a few chapters aimed not at the children but at their teachers. One of these chapters, with the title "The Imperialism of Mathematics," is a diatribe against the dominance of mathematics in the French educational system. He writes:

> Whenever an entrance examination is instituted in a scientific discipline, it invariably becomes an exercise in mathematics.... Why is there such a focus on mathematics? In reality, the trend toward mathematization turns our graduates, our future engineers, into hemiplegics....They may have learned to master certain tools, to prepare reports, but they will suffer from crippling weaknesses in observation, manual skills, common sense, and sociability.

De Gennes is not a typical French intellectual. He mixes theory with experiment, and prefers concrete objects to abstract ideas. In his research and in his teaching, he fights against the imperialism of mathematics.

In America we have the opposite situation. Our children study a variety of subjects without much formal discipline, and most of them remain mathematically illiterate. It is good for us to be reminded that different countries have profoundly different cultures and different virtues and vices. The imperialism of mathematics is difficult for Americans to imagine, but for France it is a real problem. If American children could learn more mathematics and French children less, both countries would benefit. Americans should not be misled by de Gennes's diatribe into thinking that we have nothing to learn from France. He describes eloquently the vices of the French educational establishment. He does not emphasize its virtues. The most important virtue of the French system is the strict discipline that it imposes. Every child and every student must meet rigid standards of knowledge and skill. De Gennes takes for granted the fact that the children he is talking to are literate and have a firm grasp of elementary mathematics. Americans should ask themselves why such a standard of literary and mathematical competence cannot be taken for granted in America.

Ekeland does not entirely exclude people who were not French from his narrative. He recognizes the great contributions of Galileo, Newton, Leonhard Euler, and Darwin to the development of modern science, and the great contributions of the historians Thucydides and Francesco Guicciardini to the understanding of human destiny. Some of the most illuminating passages in the book are quotations from Thucydides and Guicciardini, both of them generals who fought on the losing side in catastrophic wars and then wrote their histories to teach whatever bitter lessons posterity might learn from their defeat. Both of them saw tragedy arising not from implacable fate but from human folly and unlucky accidents. With wiser leaders, mistakes might have been avoided and tragedy averted. The worst mistakes are mistakes of overconfidence, made by arrogant leaders who do not respect the skill of their enemies or the vagaries of chance. For the

American edition of his book, Ekeland has inserted some acid remarks about arrogance and overconfidence displayed in recent actions of the American government.

A different book about the cultural history of the last four hundred years might have been written by a different Ekeland who was educated in the Anglo-American tradition instead of the French. I call the imaginary Ekeland Akeland, and I assume that Akeland is as strongly biased toward English as Ekeland is biased toward French. For Akeland, modern science still begins with Galileo, but then continues with Francis Bacon instead of Descartes. Bacon was three years older than Galileo and thirty-five years older than Descartes. Bacon pushed English science as strongly in the direction of experiment as Descartes pushed French science in the direction of theory. Bacon had a low opinion of theory. He wrote: "The logic now in use serves rather to fix and give stability to the errors which have their foundation in commonly received notions than to help the search after truth." Bacon preached humility toward nature as the only way to arrive at truth: "Man, being the servant and interpreter of Nature, can do and understand so much and so much only as he has observed in fact or in thought of the course of Nature; beyond this he neither knows anything nor can do anything." He had a grand vision of the future of science but a modest view of the science of his own time: "For though it be true that I am principally in pursuit of works and the active department of the sciences, yet I wait for harvest-time, and do not attempt to mow the moss or to reap the green corn." He did not live to see the harvest of discoveries that began thirty-four years after his death when the Royal Society of London was founded. He died while the corn was still green and Descartes had not yet started to mow the moss.

In Akeland's version of history, the scientist who personifies eighteenth-century enlightenment is Benjamin Franklin rather than

Maupertuis. Instead of the mathematicians Lagrange and Poincaré, the scientists who bring us into the modern world are the nineteenth-century British physicists Michael Faraday and James Clerk Maxwell, who set out the basic laws of electricity and magnetism. Bacon, Franklin, Faraday, and Maxwell, the chief characters in Akeland's narrative, are nowhere mentioned by Ekeland. Likewise, Akeland fails to mention Descartes, Maupertuis, Lagrange, and Poincaré. His main theme is the emergence of electricity in the eighteenth century as the growing point of science. Electricity was a product of purely Baconian science, emerging from unexpected observations of nature rather than from mathematical deduction.

Ekeland's book puts mathematical optimization at the focus of history. Optimization means choosing the best out of a set of alternatives. Mathematical optimization means using mathematics to make the choice. Maupertuis is the central character of the history because he claimed that the universe is mathematically optimized. Akeland's book has the opposite emphasis. For Akeland, things are more important than theorems. Experiments are more important than mathematics. The great scientific achievement of the Enlightenment was the experimental study of electricity. Electricity was the driving force of science for two hundred years, from the death of Newton to the rise of molecular biology. Electricity also enlarged the scope of science, moving out from the logical and mechanical universe of Newton into the color and variety of the modern world. The biologist Stephen Jay Gould formulated the philosophical principle that Akeland borrows for the title of his book: "We are the offspring of history and must establish our own paths in this most diverse and interesting of conceivable universes." Instead of mathematical optimization, Akeland postulates maximum diversity as the governing principle of the universe. His title is *The Most Interesting of All Possible Worlds: Electricity and Destiny*.

Franklin had no theoretical understanding of electricity. Electricity was outside the Newtonian domain of mechanics and gravitation that constituted the theoretical science of his time. Franklin explored electricity because it was a part of nature that nobody understood. Without pretending to understand electricity, he learned how to control it. His invention of the lightning conductor made him world-famous and earned him a warm welcome when he came to live in France. He came to France too late to meet with Maupertuis. If they had met, they would have found that they had much in common. Franklin was only eight years younger than Maupertuis. Both were good organizers as well as good scientists. Franklin was organizing the American Philosophical Society in Philadelphia while Maupertuis was organizing the Prussian Academy in Berlin. Both were gentlemen of the Enlightenment, adventurers and travelers in an age when travel was slow and arduous. Both were by temperament optimists, but neither was a Pangloss. The only serious difference between them was that Maupertuis was a mathematician and Franklin was an experimenter.

The next pair of characters in Ekeland's and Akeland's histories were Lagrange in France and Faraday in England. They lived in different centuries and had less in common than Maupertuis and Franklin. They were extreme examples of Cartesian and Baconian scientists. Faraday explored the new worlds of electricity and magnetism, chemistry and metallurgy, pushing into unknown territory far ahead of any theoretical understanding. Lagrange (1736–1813) created the science of analytical mechanics, an abstract mathematical framework that included all the results of Newtonian dynamics as special cases. Each was master of his trade, but theirs were very different trades. By unifying Newton's ideas into a single scheme, Lagrange left the world simpler than he found it. By discovering a host of unexpected new phenomena, Faraday (1791–1867) left the world more complicated than he found it. Lagrange was a unifier; Faraday was a diversifier. Although Lagrange's great work was published three years before Faraday was born, Faraday never read it and never felt a need for it. All the mathematics that Faraday needed was elementary arithmetic and a little algebra.

The histories of Ekeland and Akeland begin to diverge with Maupertuis and Franklin and reach a point of maximum divergence with Lagrange and Faraday. With the last pair of characters, Poincaré and Maxwell, the histories converge. Poincaré (1854–1912) was a mathematician with a taste for diversity. He was interested in the new science of electromagnetism as well as the old science of mechanics, and he discovered in the dynamics of stars and planets a variety of chaotic motions that Lagrange never dreamed of. Maxwell (1831–1879) was a physicist with a passion for unification. Starting from the observations of Faraday, he discovered the equations that unify the theories of electricity and magnetism and light into a mathematical structure as elegant as Lagrange's mechanics. The convergence of Ekeland and Akeland became complete when Poincaré explored the group of symmetries of the Maxwell equations, the group that is now known to physicists as the Poincaré Group. Maxwell and Poincaré together prepared the way that led Einstein to the new world of relativity.

The real Ekeland and the fictitious Akeland are teaching us a simple lesson. Each of them gives us a slanted and partial view of history. The true history of modern science must include both of them. Modern science started its rapid growth in the seventeenth century, taking its aims and methods not from Descartes alone and not from Bacon alone but from the cross-fertilization of Cartesian and Baconian ideas. Isaac Newton, the greatest figure in the history of the physical sciences, was an intimate mixture of Descartes and Bacon. He was Baconian in his study of optics, when he separated white light into its colored components and invented his reflecting telescope. He was Cartesian when he wrote his *Principia Mathematica*, deducing the system of the world from a logical sequence of mathematical propositions. He cleverly used a Cartesian style of argument, together with a Baconian knowledge of planetary motions, to demolish Descartes's cosmology of vortices in space.

In a true history of science, mathematics and electricity make equal contributions to human destiny. Our world may be the best of all possible worlds and may be the most interesting. Both possibilities are open. Our destiny depends on choices that we have not yet made, probably concerned more with biology—and particularly with our incipient understanding of the human brain—than with mathematics or electricity.

3

ROCKET MAN

IN THE SUMMER of 1944, the population of London was accustomed to the loud rumbling of a buzz bomb flying overhead, the abrupt silence when the engine stopped and the bomb began its descent to earth, the anxious seconds of waiting for the explosion. Buzz bombs, otherwise known as V-1s, were simple pilotless airplanes, launched from sites along the French and Dutch coasts. As the summer ended and our armies drove the Germans out of France, the buzz bombs stopped coming. They were replaced by a much less disturbing instrument of murder, the V-2 rockets launched from more distant sites in western Holland. The V-2 was not nerve-wracking like the buzz bomb. When a V-2 came down, we heard the explosion first and the supersonic scream of the descending rocket afterward. As soon as we heard the explosion, we knew that it had missed us. The buzz bombs and the V-2 rockets killed a few thousand people in London, but they hardly disrupted our civilian activities and had no effect at all on the war that was then raging in France and in Poland. The rockets had even less effect than the buzz bombs.

To me at that time the V-2 rockets were a cause for joy and wonder. I was a civilian scientist analyzing the causes of bomber losses for the Royal Air Force Bomber Command. I knew that the main cause of our bomber losses was German fighters, and I knew that the Germans were desperately short of fighters. If the Germans had had five times as many fighters, they could have stopped us from flying over Germany, and that would have made it much harder for us to invade their country and finish the war. I knew that the buzz bomb was a cheap and simple device but the V-2 was complicated and expensive. Each V-2 cost the Germans at least as much in skilled labor and materials as a modern fighter aircraft. It was incomprehensible to me that the Germans had chosen to put their limited resources into militarily useless rockets instead of crucially needed fighters. Each time I heard a V-2 explode, I counted it as one German fighter thrown away and ten fewer of our bombers downed. It seemed that some unknown benefactor in Germany was unilaterally disarming the German air force for our benefit. I had no idea then who the benefactor might be. We now know his name. It was Wernher von Braun.

Michael J. Neufeld's *Von Braun: Dreamer of Space, Engineer of War* is a meticulously researched and technically accurate biography.* Von Braun was not intentionally working for Germany's enemies in 1944. He was at that time a patriotic German, working for the fatherland, producing V-2 rockets for the German army. It was not his fault that V-2 rockets were not what the Germans needed for defending the fatherland. He was our benefactor only by accident. Von Braun's primary purpose, from the time he began rocket experiments as an amateur at the age of eighteen until the end of his life, was interplanetary space travel.

In 1932 he was recruited by the German army to develop rockets for military missiles. The army gave him what he wanted: steady funding and freedom to experiment. He pushed hard to develop a

*Knopf/Smithsonian National Air and Space Museum, 2007.

rocket that could fly into space, not caring whether or not the army had a reasonable military mission for it. The result of his pushing was the V-2, the first long-range ballistic missile, capable of delivering a one-ton explosive payload with very poor accuracy to a range of two hundred miles. When the V-2 made its first successful flight in October 1942, this was a big step toward von Braun's dream of walking on Mars. It should have been obvious to German military and political leaders that it was, from a military point of view, an expensive and useless toy.

How did it happen that Hitler gave his blessing to a crash program to produce the V-2 in quantity? Hitler was not a fool. As a foot soldier in World War I he had survived some heavy artillery bombardments. Von Braun demonstrated his plans for the V-2 to Hitler in person in August 1941, and Hitler reacted with sensible objections. He asked whether von Braun had worried about the timing of the explosion, since a normal artillery shell arriving at supersonic speed would bury itself in the ground before exploding and do little damage. This was a serious problem, and von Braun had to admit that he had not thought about it. Hitler then remarked that the V-2 was only an artillery shell with a longer range than usual, and the army would need hundreds of thousands rather than thousands of such shells in order to use them effectively. Von Braun agreed that this was true.

After the session with von Braun, Hitler ordered the army to plan production of hundreds of thousands of V-2s per year, but not to begin production until the bird had successfully flown. This decision seemed harmless at the time, but it played into the hands of the army rocketeers. The army leaders knew that the notion of producing hundreds of thousands of V-2s per year was absurd, but they accepted the order. It gave them authority to spend as much as they wanted on the program, without any fixed timetable. In August 1941 the war was going well for Germany. The army had won huge victories in the first two months of the Russian campaign, France was knocked out of the war, and America was not yet in. Hitler did not imagine that within three years he would be fighting a defensive war for the survival of the Reich. He did not ask whether the V-2 might be a toy that the Reich could not afford.

In Germany as in other countries, the main factor driving acquisition of weapons was interservice rivalry. The army wanted the V-2 because of rivalry with the Luftwaffe. The German air force was leading the world in high-technology weapons, developing jet aircraft and rocket aircraft and a variety of guided rocket missiles. The army had to have a high-technology project too. The V-2 was a high-technology version of artillery. It gave the army the chance to say to the air force: our rockets are bigger than your rockets.

Although Hitler was nominally a dictator, he was no more successful than political leaders of democratic countries in keeping rivalries between different branches of the military under control. He could fire military leaders, and did so from time to time, but he could not make them do what he wanted. The army leaders, with the help of von Braun, launched a crash program to produce the V-2. They produced a few thousand V-2s altogether, enough to outshine the air force but not enough to be militarily useful. Hitler could not force them to produce as many as he thought necessary, and he could not force them to stop the program and transfer its resources to the air force. The army and the air force continued to operate as independent principalities until the day Hitler died.

Von Braun's career as a rocket-builder was divided into six periods in which he worked for six different masters. From age eighteen to twenty, he worked as an amateur in Berlin with the Verein für Raumschiffahrt, the Society for Space Travel, a private group of rocket enthusiasts. He was technically the most competent member of the group. In the years 1930 to 1932 he built and successfully launched at a small airfield near Berlin a series of liquid-fueled rockets. Rockets are of two kinds: solid-fueled and liquid-fueled. Both kinds are driven forward by hot gas escaping from the back when the fuel burns. Solid-fueled rockets are simpler and cheaper. They were used unsuccessfully by the British navy attacking Fort McHenry in 1814, as recorded in the US national anthem. Liquid-fueled rockets fly faster and farther, but are much more complicated and difficult to handle.

From age twenty to twenty-eight von Braun worked as a civilian for the German army. The army acquired a large area of land at Peenemünde on the Baltic coast of Germany, and built facilities there for the large-scale development and testing of rockets. Von Braun's mother had lived nearby as a child and suggested the place as suitable for her son's activities. Von Braun's friend Walter Dornberger, an army major, was in charge of the program. Von Braun served under him as technical director of the Peenemünde establishment.

From age twenty-eight to thirty-three, during the years of World War II, von Braun continued to work at Peenemünde as a civilian for the German army, but he was legally an officer in the SS. This meant that he was under SS discipline. He wore his SS uniform as little as possible, and only on formal occasions. He disliked and distrusted his SS colleagues. But when, toward the end of the war, the SS took over the manufacture of V-2 missiles from the army, he had to do what the SS ordered. During the final weeks of the war, when he was evacuated with the remnants of the Peenemünde staff to the southeast corner of Germany, he was escorted by SS guards to keep him in line.

From age thirty-three to forty-eight he worked for the US Army at El Paso, Texas, and Huntsville, Alabama, as the leader of a large group of German rocket experts. These experts were hastily recruited in 1945 by the US forces occupying Germany to keep them out of Soviet hands, transferred to the United States, and then employed in developing Redstone missiles for the army. From age forty-eight to sixty, von Braun worked for the newly created NASA, first at Huntsville and later in Washington. The Army Ballistic Missile Agency at Huntsville became the NASA Marshall Space Flight Center in 1960, with von Braun in charge of the development of the huge Saturn booster rockets that safely carried twenty-one Apollo astronauts to the moon and back. From age sixty until his death at sixty-five, he worked for Fairchild Industries in Washington. At Fairchild he worked as hard as ever, supervising a variety of technical projects, helping to develop new airplanes and satellites for military and civilian missions.

The central concern of Neufeld's book is the third period of von Braun's life, the five years during World War II in which he realized his dream of shooting rockets into space and accepted a position of responsibility in the SS. The SS was the most criminal part of the Hitler regime, directly responsible for the administration of the concentration camps in which millions of prisoners were either murdered, starved to death, or used as slave laborers. Von Braun knew at first hand the dark side of the SS. After the Peenemünde complex was seriously damaged by an RAF bombing attack in 1943, the SS took over the production of V-2 rockets, and the main production line was moved to an underground factory called Mittelwerk that would be safe from air attacks. Mittelwerk was conveniently located near the Dora concentration camp and the town of Nordhausen in central Germany. Dora prisoners became a large part of the workforce at Mittelwerk, with SS guards to control them. Thousands of prisoners were confined in the tunnels where they worked under horrible conditions and slept on straw or bare rock. A large number of them died of hunger and disease. A smaller number were publicly hanged for disobedience or alleged acts of sabotage.

The boss at Mittelwerk was an SS general called Hans Kammler whom von Braun feared and hated. Von Braun was not responsible for running the operations. He was only a technical adviser. But he visited Mittelwerk many times to supervise the production process and improve the quality of the output. The facts about von Braun's activities at Mittelwerk and his SS membership were first revealed in a book, *Geheimnis von Huntsville* (The Secret of Huntsville), by Julius Mader, published in East Berlin in 1963. This book was not translated into English and attracted little attention in the US, being dismissed as Communist propaganda. A later book, *Dora* by Jean Michel, originally written in French but published in English in 1979, reported the same facts and attracted much more attention. Neufeld's book contains nothing essentially new but adds many details that the author found in unpublished papers by von Braun and others. Von Braun must have been well aware of the atrocities being committed in the tunnels, even if he avoided personal contact with the prisoners.

Von Braun was never interested in Nazi ideology. He belonged to the old aristocratic class of the Prussian nobility who owned big estates in Pomerania or Silesia, now annexed by Poland, or in East Prussia, now annexed by Russia. His father's estate was in Silesia, his mother's in Pomerania. These were the people who ruled Prussia for hundreds of years and ruled Germany from 1871 to 1918. They were for the most part highly educated and capable administrators, conscientious public servants, and social snobs, having more in common with their aristocratic cousins in other European countries than with the common people of Germany. They despised the socialist riffraff who came to power in 1918 and established the Weimar Republic.

They despised equally the Nazi riffraff who destroyed the republic in 1933 and gave supreme power to Hitler. But they respected Hitler as an effective leader who brought order and prosperity to Germany after the chaos and misery of the Weimar years. Hitler was, after all, more nationalist than socialist. He did not threaten their social position or their estates. Most of them served him willingly as the leader of Germany, while continuing to despise the Nazis as social and intellectual inferiors.

Wernher's father was a typical member of the Prussian nobility. He spoke three languages fluently and his wife spoke six. His three sons grew up in Berlin, in an atmosphere of wealth and privilege. Born in 1912, Wernher was sent to a private boarding school in Ettersburg Castle near Weimar with high intellectual standards and high fees. His friends there were boys of his own class. At school he became obsessed with rockets. He read the classic text *The Rocket into Interplanetary Space*, published by the rocket pioneer Hermann Oberth in 1923. He decided that his mission in life was to bring Oberth's dreams to reality. At age thirteen he made a good start by studying the mathematics that he needed in order to understand Oberth's equations. At age sixteen he became a member of the Society for Space Travel. At age eighteen, when he graduated from school, he was proficient enough in the theory and practice of rocketry to become the society's chief experimenter.

Von Braun did not hesitate to accept the job of military rocket developer that the army offered to him in 1932. Hitler was not yet in power, and the army was a conservative institution. It was interested in unmanned missiles rather than manned spaceships, but the same rockets that would drive missiles could later be used to drive spaceships. He found the army rocket people congenial. They were unpolitical like himself, good at working together on difficult technical problems and staying out of the limelight. When Hitler became chancellor in 1933, nothing much changed for von Braun. The army remained unpolitical, and the budget for rocketry continued to grow.

Change came in 1939 when Germany went to war, the army rockets were no longer technical toys but real weapons, and the SS tried to take over the program. The decisive moral choice for von Braun came in 1940, when he was asked by the army to become an SS officer. He did not want to have anything to do with the SS, so he went to his superior officer, Dornberger, for advice. Dornberger told him there were only two alternatives: either he must accept the SS commission or he could no longer work with the army. This had been decided at a higher level in the government. Von Braun would not abandon the army project to which he had devoted eight years of his life, so he said yes to the SS.

One of his friends in the project expressed dismay when he appeared in an SS uniform. Von Braun told him unhappily, "*Es geht nicht anders*" ("There is no other way"). There was another way that von Braun might have taken: to give up his dreams of rocketry and volunteer for service to his country as a soldier or an airman. He was a trained pilot and loved flying, so he might have enlisted in the Luftwaffe and served the fatherland by shooting down RAF bombers. But his dislike of the SS was not strong enough to make that other way seem reasonable.

On February 21, 1944, came von Braun's moment of partial redemption, when he stood firm against the devil to whom he had sold his soul. He was unexpectedly summoned to a private meeting with Heinrich Himmler, the chief of the SS and the second-most-powerful man in Germany. By this time the V-2 was supposed to be ready for operational use against England but was delayed by technical problems. Himmler invited him to stop working for the army and move over to the SS, bringing the entire rocket program with him. Von Braun reported the conversation in a memoir written six years later.* Himmler said:

---

*The original version of the 1950 memoir was unpublished, and is now in the von Braun papers at Huntsville. A revised version was published with the title "Reminiscences of German Rocketry" in *Journal of the British Interplanetary Society*, Vol. 15 (1956). The memoir is historically unreliable, written for an American or British audience long after the event. No independent report of the conversation with Himmler exists.

> Why don't you come to us? You know that the Führer's door is open to me at any time, don't you? I shall be in a much better position to help you lick the remaining difficulties than that clumsy Army machine!

Von Braun politely declined the invitation. According to his memoir, he ventured to compare the V-2 with "a little flower that needs sunshine, fertile soil, and some gardener's tending." He told Himmler that "by pouring a big jet of liquid manure on that little flower, in order to have it grow faster, he might kill it." His reason for refusing the invitation was probably concern for the welfare of his beloved rockets rather than concern for the welfare of the Dora prisoners. Still it took courage to refuse an invitation from Himmler. It took even more courage to compare the help offered by the chief of the SS to a load of shit.

"One month later, the pay-off came, Himmler-style," von Braun reported in his memoir. Gestapo agents knocked on his door in the middle of the night and took him to a prison cell in Stettin on the Baltic coast in present-day Poland. After a week in the cell, he was given a hearing before three SS officers and formally accused of sabotaging rocket development, making defeatist remarks about the war, and planning to fly to England with all the plans for the V-2. Meanwhile, with the help of the armaments minister Albert Speer, who was a personal friend both of von Braun and of Hitler, Dornberger succeeded in obtaining a piece of paper signed at the Führer's headquarters, releasing von Braun provisionally for three months. Von Braun sat in jail for only ten days and was not physically abused. Those ten days were of enormous value to him when he came to the United States. Whenever people asked him about his past, he could mention those days as evidence that he had not been a Nazi. He never claimed that he had actively resisted the Nazi regime, but the story of his imprisonment made him appear to have been a victim of the Nazis rather than an accessory to their crimes.

The second half of Neufeld's book describes von Braun's life in America after 1945. He adapted with astonishing speed to the American way of life. In 1946 he became a born-again Christian and joined the congregation of a small Church of the Nazarene in Texas. For several years he worked patiently for the army, refurbishing surplus V-2 rockets that the US had imported from Germany. The army could not give him more interesting work because there was no money for further development of rockets. He quickly understood that in America the money was controlled by Congress and Congress was controlled by public opinion. The money was lacking because the public was not interested in rocketry. So he resolved to go directly to the public.

Whenever he had the chance, first with magazine articles and then with speeches on radio and television, he preached the gospel of rocketry. He spoke not only about unmanned rockets to defend the country but about manned rockets to explore the solar system. It took him only seven years from his arrival in the United States to become world-famous as the chief promoter of space travel. In 1952, *Collier's* magazine published a flamboyant article with pictures of winged spaceships in orbit and a text, "Crossing the Last Frontier," by von Braun. In the next year his book *The Mars Project*, with detailed specification of rocket weights and payloads required for a manned exploration of Mars, was published in English and in German. As his fame grew, so did the budgets for the army rocket program at Huntsville.

There were two high points of von Braun's life in America. In 1958, after the Soviet Union launched Sputnik and the US Navy Vanguard satellite crashed ignominiously on its launchpad, von Braun's team at Huntsville successfully put *Explorer 1*, the first American satellite, into orbit. In 1969, he watched Neil Armstrong and Buzz Aldrin walk on the moon, carried there by his rockets and fulfilling his dream of the human race moving out of the nursery. Von Braun was unique as an organizer of big projects who could persuade prima donnas to work harmoniously together, and who also understood every detail of the hardware.

After 1969, he remained as busy as ever, but his hopes for going on to Mars faded. Five more Apollo missions reached the moon successfully, and one, *Apollo 13*, was an epic failure from which the crew came home safely. After that, the public was not interested in going further. Budgets rapidly decreased and the Apollo program ended. All that von Braun could do to keep manned rocket missions alive was to promote the Space Shuttle, a reusable ferry vehicle that had originally been the bottom part of his Mars Project. The shuttle was supposed to be cheap and safe, flying frequently with a quick turnaround between missions. When after many delays the shuttle finally flew, it turned out to be neither cheap nor safe nor quick. He was lucky not to live long enough to see how miserably the shuttle would fail.

This book raises three important issues: one historical and two moral. The historical question is whether von Braun's great achievement, providing the means for twelve men to walk on the moon, made sense. Was it a big step toward the realization of his dream of colonizing the universe, or was it a dead end without any useful consequences? In the short run, the Apollo program was certainly a dead end. As a public program dependent on the taxpayers' money, it collapsed as soon as the taxpayers lost interest in it. When von Braun moved from NASA to Fairchild Industries in 1972, he was wagering that human adventures in space would in the future be better supported by private investors than by governments. He died of cancer five years later. Now, thirty years after his death, we see a vigorous growth of privately funded space ventures. If von Braun had lived twenty years longer, he might have pushed us sooner into the era of private launchers. He might even have rescued the Space Shuttle, his orphaned baby, and made it become what he had intended it to be: cheap and safe and quick. In the long run, one way or another, people will again dream of colonizing the universe and will again build spaceships to embark on celestial journeys. When that happens, they will be following in von Braun's footsteps.

The two moral issues that Neufeld's book raises are whether von Braun was justified in selling his soul to Himmler and whether the United States was justified in giving sanctuary and honorable employment to von Braun and other members of the Peenemünde team. Some of the other scientists at Peenemünde were guilty of worse offenses than von Braun. The most notorious was Arthur Rudolph, a close friend of von Braun, who had been an enthusiastic Nazi and served as the chief of production at the Mittelwerk factory. Rudolph was far more directly involved than von Braun in the exploitation and abuse of prisoners. After that, Rudolph lived in the United States for thirty-nine years and enjoyed a distinguished career as a rocket engineer. Finally, in 1984, formerly secret documents describing Rudolph's activities in Germany emerged into the light of day, and he was threatened with a lawsuit challenging his right to American citizenship. Rather than fighting the lawsuit, he renounced his citizenship and returned with his wife to Germany. One of the investigators of the Rudolph case said, "We're lucky von Braun isn't alive." Von Braun had died, full of years and honor, seven years earlier. If von Braun had been alive in 1984, with his public fame and political clout intact, he would have come to the defense of Rudolph and probably won the case.

Neufeld condemns von Braun for his collaboration with the SS, and condemns the US government for covering up the evidence of his collaboration. Here I beg to differ with the author. War is an inherently immoral activity. Even the best of wars involves crimes and atrocities, and every citizen who takes part in war is to some extent collaborating with criminals. I should here declare my own interest in this debate. In my work for the RAF Bomber Command, I was collaborating with people who planned the destruction of Dresden in February 1945, a notorious calamity in which many thousands of innocent civilians were burned to death. If we had lost the war, those responsible might have been condemned as war criminals, and I might have been found guilty of collaborating with them.

After this declaration of personal involvement, let me state my conclusion. In my opinion, the moral imperative at the end of every war is reconciliation. Without reconciliation there can be no real peace. Reconciliation means amnesty. It is allowable to execute the worst war criminals, with or without a legal trial, provided that this is done quickly, while the passions of war are still raging. After the executions are done, there should be no more hunting for criminals and collaborators. In order to make a lasting peace, we must learn to live with our enemies and forgive their crimes. Amnesty means that we are all equal before the law. Amnesty is not easy and not fair, but it is a moral necessity, because the alternative is an unending cycle of hatred and revenge. South Africa has set us a good example, showing how it can be done.

In the end, I admire von Braun for using his God-given talents to achieve his visions, even when this required him to make a pact with the devil. He bent Hitler and Himmler to his purposes more than they bent him to theirs. And I admire the United States Army for giving him a second chance to pursue his dreams. In the end, the amnesty given to him by the United States did far more than a strict accounting of his misdeeds could have done to redeem his soul and to fulfill his destiny.

Note added in 2014: This review provoked a record number of eloquent and moving responses from people outraged by my friendly portrayal of von Braun. Here is an extract from one of them:

> I was a medical student at the London Hospital, in 1944, when an early V-2 landed one afternoon in Petticoat Lane, a crowded and popular people's market in London's East End. There were hundreds of killed and injured and over two hundred were admitted to the hospital, where the severely injured were promptly triaged to the operating rooms but many lay for hours in the corridors and basement to receive treatment, mostly for nasty lacerations from flying glass. It was a scene I have never forgotten.
>
> Professor Dyson's role in the planning of the RAF raid on Dresden, admittedly a horrific incident, seems paltry compared to the calculated killing and brutal exploitation of the inmates of the forced labor camp where the V-2 was conceived and manufactured. Von Braun never publicly renounced his role in the Nazi regime, of whose sadism and brutality he seems to have been fully aware.
>
> Surely confession and penitence must precede reconciliation? Amnesty yes, reconciliation maybe, but forgiveness no. Neither did we need to reward such a man with a presidential medal for his acts of redemption for unforgivable sins.
>
> Bernard Lytton
> Donald Guthrie Professor Emeritus of Surgery/Urology
> Yale University School of Medicine
> Director, Koerner Center for Emeritus Faculty, Yale University
> New Haven, Connecticut

In response to a letter from Leo Blitz in Berkeley, whose mother survived the concentration camp at Stutthof, I wrote:

I once visited the camp at Stutthof when I was in Poland. I am not saying that von Braun or anyone else was innocent. But I think you miss the main point. Amnesty is not for the innocent. Amnesty is for the guilty. We need amnesty at the end of a war because a large number of people on both sides are guilty. War is like that. Modern war is a brutal business, and when I was working for Bomber Command I was in the same business as von Braun. After that, we all needed an amnesty, with a few exceptions such as your mother.

4

THE DREAM OF SCIENTIFIC BROTHERHOOD

GROWING UP AS a child in England, I absorbed at an early age the notion that different countries had different skills. The Germans had Bach and Beethoven, the Spanish had Velázquez and El Greco, the French had Monet and Gauguin, and we had Newton and Darwin. Science was the thing the English were good at. This notion was reinforced when I began to read children's books of that period, glorifying the achievements of our national heroes, Faraday and Maxwell and Rutherford.

Ernest Rutherford, the New Zealander who had discovered the atomic nucleus and created the science that came to be called nuclear physics, was then at the height of his fame. Although he had immigrated from New Zealand, Rutherford became more English than the English. He spoke for England in a famous statement contrasting the continental European style with the English style in science: "they play games with their symbols, but we, in the Cavendish, turn out the real solid facts of Nature." The French and Germans were doing calculations with the abstract mathematical equations of quantum theory, while Rutherford was banging one nucleus against another and transmuting nitrogen into oxygen. English children learned to be proud of Rutherford, just as we were proud of our military heroes

Nelson and Wellington, who had beaten Napoleon. Patriotic pride of this sort is in some ways healthy. It encourages children to be ambitious and to tackle big problems. But it is harmful when it leads them to believe that they have a natural right to rule the world.

I still remember some of the patriotic poems that I had to learn by heart and recite as a seven-year-old:

Of Nelson and the North
Sing the glorious day's renown,
When to battle fierce came forth
All the might of Denmark's crown.

The battle that Nelson fought in the harbor of Copenhagen was especially famous because his commanding officer put up a flag signal ordering him to cease fire. Nelson pointed his telescope at the flag signal and looked through it with his blind eye. Since he did not see the flag, he continued the battle and won a glorious victory. But even a seven-year-old understands that Nelson's defeat of the Danes at Copenhagen was not as glorious as his defeat of the French fleet at Trafalgar four years later. Even a seven-year-old may feel some sympathy for the defeated Danes, and may question whether Nelson's undoubted bravery and brilliance gave him the right to bombard their homes. I recently visited a tavern in Copenhagen where the tourist is proudly informed that this is one of the few buildings along the waterfront that were not demolished by Nelson's guns. The collateral damage resulting from his victory is not forgotten.

John Gribbin's book *The Fellowship* belongs to the harmless kind of patriotic literature.* It is a portrait gallery displaying a group of

---

**The Fellowship: Gilbert, Bacon, Wren, Newton, and the Story of a Scientific Revolution* (Overlook, 2008).

remarkable characters who made important contributions to the rise of modern science in the seventeenth century. Each of the biographies is dramatic. Those characters lived through turbulent times, and their personal lives were as exciting as their ideas. Almost all of them are English. Gribbin is not writing a history of science but only a history of a particular institution, the Royal Society of London. "The Fellowship" means the group of men who founded the society in 1660 and devoted their time and energy to its activities. Although they were English, their aims and purposes were international, and they welcomed distinguished scholars from many countries as fellows of the society. From the beginning, one of the main activities of the society was the exchange of information and the improvement of contact between England and the rest of the world. The founding of the society was not the beginning of modern science, but it was a unique event with great consequences, well worth studying in detail. Gribbin's book gives a lively and readable account of it.

The story begins a hundred years earlier with William Gilbert, a medical doctor who practiced in Colchester and London and became president of the Royal College of Physicians in 1600. He was one of the royal physicians responsible for keeping Queen Elizabeth in good health. In his spare time he did experiments on magnetism and published his conclusions in a book with the Latin title *De Magnete*. The full title in English is *On the Magnet, Magnetic Bodies, and the Great Magnet the Earth: A New Physiology Demonstrated by Arguments and Experiments*. The book is written in a remarkably modern style, putting the science of magnetism on a firm experimental foundation. Gilbert did careful measurements, mostly using as his experimental material little spheres of natural lodestone (magnetic oxide of iron), which he called *terrellae*, or in English "little earths." He was aware from the beginning that these little magnets were models for the earth. He suspended them in water and measured their attractions and repulsions in detail. He cleared up a great deal of confusion by demonstrating that the use of the words "North Pole," to mean the end of a magnet that pointed north, was wrong. He demonstrated that north and south poles attract each other, and therefore, if the magnet were taken to be a model of the earth, the end of the magnet that pointed north would correspond to the south pole of the earth. He says in his book:

> All who hitherto have written about the poles of the loadstone, all instrument-makers, and navigators, are egregiously mistaken in taking for the north pole of the loadstone the part of the stone that inclines to the north, and for the south pole the part that looks to the south: this we will hereafter prove to be an error.

Roughly speaking, Gilbert did for the science of magnetism the same job that Benjamin Franklin did for the science of electricity two hundred years later, establishing the basic facts by means of experiments that anyone who doubted his conclusions could repeat. But Gilbert, since he lived two hundred years earlier, was in some ways the greater pioneer. In the course of his study of magnets, he also did a number of experiments on electricity, demonstrating that electric and magnetic materials were different and should be studied separately. Gilbert was aware that he was pioneering a new style of experimental philosophy that could be extended to many other subjects besides magnetism. He writes in the preface to *De Magnete*:

> To you alone, true philosophers, ingenuous minds, who not only in books but in things themselves look for knowledge, have I dedicated these foundations of magnetic science—a new style of philosophizing.

One of the people who read *De Magnete*, probably soon after it appeared in 1600, was Galileo. Galileo was twenty years younger than Gilbert, but already well started in his studies of dynamics, using pendulums and balls rolling down inclined planes as his experimental tools. Galileo in his correspondence with friends wrote warmly of Gilbert: "I greatly praise, admire and envy this author, that a conception so stupendous should have come to his mind." Galileo later did experiments himself with magnets and confirmed Gilbert's results. Fortunately, the friendly relations between Galileo and his English admirers were not disturbed by disputes over priority of the kind that arose between Newton and Leibniz a century later. Gilbert was given some share of the glory that Galileo earned as the father of modern experimental science.

After Gilbert and Galileo comes Francis Bacon, who, unlike the other characters in the story, seldom did an experiment. He was a man of many talents, so gifted that he was seriously proposed in later centuries as the author of Shakespeare's plays. At the age of fifteen he was helping the English ambassador in Paris with diplomatic correspondence, and developed a serious interest in codes and cryptography. He later became a successful writer, lawyer, and politician. He was lord chancellor in 1618 and was disgraced for taking bribes in 1621. After his disgrace, he spent five years in retirement writing fragments of a great work that remained unfinished, *The Great Instauration*. By "instauration" he meant an organization for acquiring knowledge from all over the world and putting it to practical use.

The essential feature of his vision was that the increase of knowledge should be a collective activity, with organized groups of people observing in detail how nature works. After the observations were collected, another group of people, scholars and philosophers, would interpret the results and deduce the laws that nature follows. Finally, a third group of people, inventors and manufacturers, would use their knowledge of nature's laws for the advancement of human wealth and welfare. This blueprint for the building of a knowledge-based society was very far ahead of its time. In many ways, Bacon's scheme resembles the institutions of science and technology in the twenty-first century more than it resembles the Royal Society in the seventeenth. Nevertheless, the founders of the Royal Society were strongly influenced by Bacon's writings and believed that they were helping to make his dreams come true. And now, 350 years later, it turns out that they were right.

Bacon was a master of a literary form that he called the essay. His essays are brief, usually a couple of pages summarizing his views about a big subject. Many of his essays have become classics, distilling much wisdom into a few words. Here are a few of his memorable statements about the pursuit of knowledge:

> All depends on keeping the eye steadily fixed on the facts of nature, and so receiving their images as they are. For God forbid that we should give out a dream of our own imagination for a pattern of the world.
>
> Man is the helper and interpreter of Nature. He can only act and understand in so far as by working upon her or observing her he has come to perceive her order. Beyond this he has neither knowledge nor power.
>
> Truth emerges more readily from error than from confusion.
>
> The true and lawful goal of the sciences is simply this, that human life be endowed with new discoveries and powers.

After his death in 1626, his most imaginative work was published, a novel with the title *New Atlantis*, describing a utopian society living on an island in the South Pacific and directed by an organization called the Foundation. The Foundation is a group of philosophers dedicated to scientific research and human improvement:

> The End of our Foundation is the knowledge of Causes, and secret motions of things; and the enlarging of the bounds of Human Empire, to the effecting of all things possible.

Bacon died amid a chaos of unpaid debts and unfinished manuscripts. He never knew which of the many seeds that he planted would bear fruit. The New Atlantis turned out to be one of the most fertile. Thirty years after his death, the name "Fellows," which he gave to the members of his Foundation, was borrowed by the founders of the Royal Society for the members of theirs. And three hundred years later, the writer Isaac Asimov borrowed the name "Foundation" for one of the most popular series of science-fiction stories ever written.

The next of the English pioneers was William Harvey, the physician who revolutionized the practice of medicine by discovering the circulation of the blood. The title of his great work published in 1628 was *Anatomical Exercises on the Motion of the Heart and Blood in Animals*. He was trained in Padua, where he was a student of Fabricius, a famous anatomist who made careful dissections of animals and identified the valves in veins. Fabricius did not understand the function of the valves, since he believed the prevailing dogma that veins and arteries both carried blood away from the heart. After Harvey returned to England he did careful experiments, tying bandages around the arms of his patients and observing how the flow of the blood in the veins responded. He found that the function of the valves was to block flow away from the heart and allow flow toward the heart. These simple observations proved that the blood circulates through the body, away from the heart through the arteries and back to the heart through the veins. Harvey also showed that a separate circulation takes blood from the heart to the lungs and back again.

After Harvey came the "great generation," the group of about twenty people who came together in 1660 to launch the Royal Society. The main purpose of Gribbin's book is to explain how and why this happened. How did it happen that so many people with wealth and education became seriously interested in science? And why did they concentrate their attention on experiments and observations of nature rather than on philosophical theorizing? Gribbin answers these questions by examining the historical circumstances out of which this group of people arose.

The central fact about the founding of the Royal Society is that it coincided with the restoration of the English monarchy under King Charles II. England had been torn apart by civil war for nine years, from 1642 to 1651. Parliamentary forces led by Oliver Cromwell defeated royal forces led by Charles I. Charles I was beheaded in 1649 and England became a republic, governed by Cromwell as lord protector. Charles II spent nine years in humiliating exile, wandering between France and Holland and Spain. When Cromwell died in 1658, his second-in-command, General George Monck, started to talk with the defeated Royalist leaders and quickly negotiated a deal. Charles II would be invited back as king, and only a few ringleaders of the gang that had killed his father would be punished. Most English people were tired of religious squabbles. They had no wish to fight the civil war over again. So Charles II came back and successfully reunited the country, governing with a light hand and making whatever compromises were needed to stay on his throne. He reigned for twenty-five years, more or less peacefully, and before he died somebody composed a poem for his tombstone:

Here lies our sovereign lord King Charles, whose word no man relies on,
Who never said a foolish thing and never did a wise one.

Charles II had learned from his father's mistakes not to take himself or his job too seriously. That was the background against which the Royal Society came into being.

During the nine years of civil war and the years of Cromwell's rule that followed, upper-class Englishmen found themselves divided, isolated, and insecure. These were the landowners and merchants and men-about-town who were accustomed to running their estates and businesses and also to running the country. Many of them had been friends of the king, others were friends of Cromwell. Gribbin gives us a vignette of Harvey, who was a friend of Charles I in 1642 when the war began. Charles was busy leading his troops in the first serious battle of the war at Edgehill, which ended in a draw. He left his two sons in the care of Harvey. So Harvey sat under a hedge on the battlefield with the two future kings, Charles, then aged twelve, and James, aged nine. All of them survived the battle, and it is possible that young Charles acquired from Harvey some of the interest in science that he put to good use when he became king eighteen years later. But Harvey had to pay dearly for his service to the royals. When Charles I was defeated and imprisoned, the parliamentary government stripped Harvey of all his honors and privileges.

For others besides Harvey, science provided an escape from turmoil and insecurity. It provided a way for men possessing property and wealth to put their leisure to good use. It also provided a way for them to forget their differences, to come together and talk about questions having nothing to do with politics and theology. The group that eventually gave birth to the Royal Society started in Oxford in

1648 under the leadership of John Wilkins. Wilkins was an amateur astronomer and engineer with a secure base of operations as the warden of Wadham College in Oxford. He was a personal friend of Cromwell and afterward married Cromwell's sister.

He had published in 1641 a book with the title *Mercury, or the Secret and Swift Messenger: Shewing, how a Man may with Privacy and Speed communicate his Thoughts to a Friend at any Distance.* This book described a system of rapid long-range communication based on bells. Using a series of relay stations, each station containing a human bell ringer with two bells of different pitch, messages could be coded and encrypted and transmitted over long distances at the speed of sound. In Oxford he started an "experimental philosophical club," with emphasis on actually doing experiments rather than merely talking and writing. His own experiments were done with transparent beehives that he constructed so that he could observe in detail how the bees organized their activities. The two most important members of the club were Robert Boyle and Robert Hooke. Boyle built a chemical laboratory in Oxford, and worked hard to separate the kernel of truth from the encrustations of myth in the processes studied by alchemists. He described his experiments in *The Sceptical Chymist*, the first account of chemistry written from a modern point of view. Hooke came to Oxford as Boyle's paid assistant and was enormously helpful to the group, as he had a genius for building experimental apparatus that worked. He improved the performance and reliability of air pumps, pendulum clocks, and microscopes, the tools that made experimental science possible.

While the members of Wilkins's club were actively engaged in doing science in Oxford, another group of gentlemen were talking about science at Gresham College in London. The Oxford group were mostly Parliamentarians, the London group mostly Royalists. The

London group did not contain scientists of the caliber of Boyle and Hooke, but it contained serious amateurs who had good personal contact with Charles II. One of them was Sir Robert Moray, an expert in chemistry who had spent some time with the king during his exile. After the king returned, Moray helped him to build a chemical laboratory at his palace in Whitehall, where the two of them worked together doing experiments. The diarist Samuel Pepys records that he once went with Moray "into the king's laboratory under his closet; a pretty place; and there saw a great many chymical glasses and things but understood none of them." Unfortunately, history does not record which of his chemical toys the king liked to play with.

In November 1660 the time was ripe to heal the wounds of the civil war, to use the king's genuine interest in science to bring Parliamentarians and Royalists together. A meeting was called at Gresham College in London to establish a new society combining the London and Oxford groups, with Wilkins as chairman. The society was duly established, "for the Promoting of Experimentall Philosophy." At a second meeting a week later, Moray brought a message from the king officially approving the foundation. One year later, the king accepted his election as a fellow of the society. And in 1663 the society received the royal charter naming it "The Royal Society of London for Promoting Natural Knowledge." Together with the charter came the Latin motto *Nullius in Verba*, which Gribbin unfortunately mistranslates as "Nothing in Words." As any educated person in the seventeenth century would have known, the word *nullius* does not mean "nothing." It is a genitive form meaning "of no one." The motto is an abbreviated version of a well-known line from the Latin poet Horace: "*Nullius addictus iurare in verba magistri*," or in English, "Sworn to follow the words of no master." It is a radical statement, a declaration of intellectual independence. It means that the society will pay attention to facts and not to scholastic or political or ecclesiastical authorities. The king was in his personal life a libertarian, sharing the subversive spirit that the motto expressed.

Once the Royal Society was established with the king's blessing, it quickly came under enormous pressure to admit noblemen and other wealthy people with no particular competence in science. The founders resisted this pressure as best they could. They were determined to maintain the core of the society as an active group of experimental scientists, while accepting a wide periphery of inactive fellows who only came to listen to lectures and provide financial support. The core of the society survived, largely due to the exertions of Hooke, who served as the curator of experiments for the society for more than twenty years, and kept experimental programs going in many fields. After Hooke, Edmond Halley and Isaac Newton in turn took responsibility for keeping the society active. Newton served as the president for more than twenty years and missed only three of the weekly meetings. Gribbin's account ends in 1759 when Charles Messier, an astronomer at the Paris Observatory, observed Halley's comet return as Halley had predicted fifty-four years earlier. This event was recognized all over Europe as the final triumph of Newtonian physics.

Was the founding of the Royal Society a major turning point in the worldwide history of science, or was it only a local event in the parochial history of England? This is the question that Gribbin's book does not answer. He mentions that the French Academy of Sciences was founded four years after the Royal Society and performs many of the same functions. In 1660 the time was evidently ripe for science to become organized on a national scale. But the French Academy was different from the Royal Society in many ways. The French Academy was a government institution, financed and controlled by the state. The scientists who belonged to it were paid civil servants. The motto *Nullius in Verba* was not for them. The academies of science that arose later in Berlin and St. Petersburg followed the French rather than the English model. The unique feature of the Royal Society was that it attempted to perpetuate the tradition of the seventeenth-century philosophical club, accepting the king's blessing but maintaining freedom from his control. The Royal Society aimed to keep science in private hands, hoping that there would always be enough talented individuals with wealth and leisure willing to devote their lives and material resources to experimental philosophy.

Inevitably, the dream that science could remain forever a Baconian brotherhood of philanthropic explorers failed. In England, as in France and other countries, science grew rapidly and soon outgrew the resources of wealthy amateurs. In England as elsewhere, most scientists became professionals and worked in universities or government laboratories. But still, the Royal Society survived and maintained its independence, and the tradition of independent amateurs making serious contributions to knowledge also survived. In the nineteenth century, Lord Rosse, who discovered spiral galaxies at his private observatory in Ireland, and Lord Rayleigh, who discovered the inert gas argon in his private laboratory in Essex, were both amateurs, and so was Charles Darwin.

The tradition set by the Royal Society has also survived in America. When Benjamin Franklin founded the American Philosophical Society at Philadelphia in 1743, he did not apply for a royal charter, but in other respects he followed the pattern set by the Royal Society. As the main business of his society he proposed, "All philosophical Experiments that let Light into the Nature of Things, tend to increase the power of Man over Matter, and multiply the Conveniencies or Pleasures of Life." His members, like the early fellows of the Royal Society, would contribute cash to share the costs of experiments. When the US National Academy of Sciences was founded in 1863, it was given a statutory duty to advise the federal government concerning scientific questions, but it remained an independent institution with its own finances and its own administration.

Finally, at the beginning of the twenty-first century, the spirit of the early Royal Society is being revived by a new breed of technocratic billionaires in America. The most famous of the new followers of Bacon is Craig Venter, the biological entrepreneur who set up his own privately funded project to sequence the human genome in competition with the government's Human Genome Project, and beat the government team at their own game. After that, he equipped his private yacht with apparatus to collect microbes from the ocean and sequence their genomes in bulk, so that he can now sail around the globe and begin the sequencing of the entire biosphere of the planet. Venter calculates that if the technologies of collection and sequencing continue to improve as expected, it should be possible within thirty years to obtain a digital blueprint of all existing forms of life. Other members of the billionaires' club are Larry Page and Sergey Brin, the founders of the Google Corporation, who have set out to reorganize all human knowledge so that it will be accessible to everybody. They have been so successful that it is now difficult to remember how we used to live a few years ago without Google to answer our questions.

Other young billionaires are starting private enterprises to explore and exploit space, with the aim of beating NASA at its own game, just as Venter beat the National Institutes of Health. These private space ventures may fail totally. They have enormous obstacles to overcome, and they are unable to agree on any clearly defined objectives. But it is still possible that one or more of them will succeed. Then a new era of exploration will begin, similar to the era of exploration that brought ships from Europe to all parts of the world in the sixteenth century. The improvement of the art of navigation was one of the central concerns of the founders of the Royal Society. Their leader, John Wilkins, wrote a book in 1638 with the title *The Discovery of a World in the Moone*, raising the question whether a voyage to the moon might one day be feasible. If our new space venturers should ever succeed in establishing a private moon base, the founders of the Royal Society will be there with them in spirit to share the glory.

Note added in 2014: In the published review I said that Bacon never did an experiment. That statement was untrue, and I am grateful to Timothy Beecroft for pointing out the error. Griffin on page 85 quotes from a letter that Bacon wrote shortly before his death: "I was desirous to try out an experiment or two, touching the conservation and induration of bodies. As for the experiment itself, it succeeded excellently well." Bacon does not describe the experiment, but Gribbin conjectures that it involved inhaling vapors. Gribbin quotes from earlier writings of Bacon that express an interest in the inhalation of niter. Inhalation experiments were intended to be medical remedies rather than scientific investigations. Gribbin concludes that the experiment that "succeeded excellently well" may have contributed to Bacon's death a few days later.

5

WORKING FOR THE REVOLUTION

DR. JOHANNES FAUST was a real person who has an entry in the German dictionary of national biography.* He was a professional astrologer and magician who spent his time wandering from town to town in Germany during the sixteenth century, providing horoscopes and astrological advice to bishops and princes as well as to the common people. He was famous enough to come to the attention of Martin Luther, who denounced him for making a pact with the devil. Whether Faust himself claimed any acquaintance with the devil is not clear. He became a legend soon after his death, when an account of his life was published in Germany, incorporating many fanciful tales borrowed from other sources.

Less than a century later, Christopher Marlowe wrote his play *The Tragicall History of the Life and Death of Doctor Faustus*, which gave the legend a dramatic form. Marlowe's Faustus speaks the immortal lines "Was this the face that launch'd a thousand ships / And burnt the topless towers of Ilium?" when the devil introduces him to Helen of Troy, and "See, see where Christ's blood streams in the

**Neue Deutsche Biographie* (Berlin: Duncker und Humblot, 1961), Vol. 5, pp. 34–35.

firmament" when his debt to the devil comes due and he is carried off to spend eternity in Hell. Two hundred years after Marlowe, Johann Wolfgang von Goethe wrote his *Faust*, an even more famous play that became required reading for every schoolchild in the German-speaking countries of Europe. Goethe's Faust is a more complicated character than Marlowe's Faustus. At the end of Goethe's play, Faust is redeemed and his pact with the devil is broken. At the beginning of the twentieth century, *Faust* was the best-known work of German literature. In England, Marlowe was outshone by Shakespeare, but in Germany, nobody outshone Goethe.

So it happened that a bunch of bright young physicists, assembled at the Institute for Theoretical Physics in Copenhagen in 1932 for their annual Easter conference, decided to entertain their elders by performing a spoof of Goethe's *Faust*. German was then the international language of physics and the main working language at Copenhagen. Everyone at the conference was fluent in German and familiar with *Faust*. At the Easter conference in 1931 there had been a similar performance with the title *The Stolen Bacteria*, a spoof of a spy movie that had recently been playing in Copenhagen. The 1931 show was composed and directed by George Gamow, famous as a joker as well as a physicist. In late 1931 Gamow had unwisely returned to his native Russia, and the Soviet government had refused to let him leave. The job of composing and producing the 1932 show was taken over by Max Delbrück, a close friend of Gamow. Delbrück was then twenty-five years old and was soon to accept a position as an assistant to Lise Meitner in Berlin. Meitner was an experimental physicist, destined to become world-famous in 1939 for her share in the discovery of nuclear fission. Gamow's performance in 1931 had been a great success. In 1932 Delbrück rose to the occasion and produced something even better.

The founder and presiding spirit of the Copenhagen institute was

Niels Bohr, the Danish physicist who had developed the first quantum theory of atoms in 1913. By his success as a fund-raiser and administrator, as well as his outstanding intellectual and human qualities, Bohr had made his institute a world center of theoretical physics. Copenhagen was the place where the leaders of the quantum revolution in the 1920s met and argued and put it all together. Bohr was indefatigable in exploring and clarifying every detail of the new theory. In Delbrück's version of *Faust*, the role of God would be played by Felix Bloch impersonating Bohr, and the role of Mephistopheles would be played by Léon Rosenfeld impersonating Wolfgang Pauli. Bloch and Rosenfeld were young contemporaries of Delbrück.

Pauli was older. At thirty-one he was regarded by the irreverent younger generation as an elder statesman, past his prime as an original thinker but still formidable as a critic. Pauli was chosen as the model for Mephistopheles because he was famous for his sharp tongue. He was ruthless in criticizing people who did not speak or think clearly. He even dared to criticize Bohr. He was proud of the title "God's whip," which he had earned by giving tongue-lashings to people who talked nonsense. In real life, Bohr and Pauli treated each other with guarded respect, like God and Mephistopheles in Goethe's play.

The model for Faust, the central role in Goethe's play, was Paul Ehrenfest, a charismatic teacher who had settled at Leiden in the Netherlands and had propelled to greatness a succession of brilliant Dutch students. Ehrenfest was a tortured soul, at home in the comfortable old world of classical physics and feeling like an alien in the weird new world of quantum mechanics. He was fifty-one years old, five years older than Bohr, and unable to make the quantum leap that Bohr had successfully accomplished. Since Faust was also a tortured soul, it was dramatically right to give his role to Ehrenfest. But when Delbrück wrote the script, he did not know the depth of Ehrenfest's pain. Delbrück gave him the lines:

So I'm the critic, sad and misbegot.
All doubts assail me; so does every scruple;
And Pauli as the Devil himself I fear.

These lines were unintentionally cruel. They fit too well the anguish that Ehrenfest was carefully concealing from his friends. If Delbrück had known how close to the edge of despair Ehrenfest had come, he would have found a way to give the role of Faust to someone else.

In real life, Pauli and Ehrenfest were close friends, and Pauli encouraged Ehrenfest's questioning attitude toward quantum theory. But Ehrenfest still felt inadequate, left behind by the younger generation of physicists who were writing papers faster than he could read them. He wrote letters to Bohr and Einstein telling them that he was thinking of committing suicide, but the letters were never mailed. A year and a half after the *Faust* performance, he killed himself in a park in Amsterdam.

At the performance of the Delbrück version of *Faust* in 1932, no hint of impending tragedy was visible. Audience and performers alike enjoyed the show hugely. The script was full of clever inside jokes that only people familiar with Goethe's play and with the personalities of modern physics could appreciate. The audience was expert in both matters. In the front row sat Bohr, Ehrenfest, Meitner, Werner Heisenberg, Paul Dirac, and Delbrück, all of them famous physicists, and all except Meitner having roles in the play. All of them, with the possible exception of Ehrenfest, laughed at the jokes and enjoyed seeing themselves and their colleagues lampooned. All of them carried away memories of an evening that was a high point of the Copenhagen institute and of twentieth-century physics.

Delbrück preserved the script of the performance but never published it. The German text is still unpublished. Thirty years after the performance, Gamow borrowed the script from Delbrück and translated it into English with the help of his wife Barbara. The English version was finally published, with illustrations by Gamow, in his book *Thirty Years That Shook Physics.** Gamow was by that time firmly established in America as a writer of popular books about science and as the founding father of big bang cosmology.

Einstein has a minor role in the play, as a king with a retinue of trained fleas who cause considerable annoyance to the other characters. The fleas are Einstein's unified field theories, which in 1932 were already becoming an obsession. His distrust of quantum mechanics, and his addiction to unified field theories, had the effect of cutting him off from his old friends. Delbrück was holding up the mirror to Einstein, to show him how he looked to the younger generation. But Einstein was not looking into the mirror. He was not in the audience.

Gino Segrè, a professor of physics and astronomy at the University of Pennsylvania, has used the Copenhagen performance in 1932 as the centerpiece for his book *Faust in Copenhagen: A Struggle for the Soul of Physics.*† The book is a history of the quantum revolution that started with a daring proposal by Max Planck in 1900. Planck suggested that light and heat radiation are emitted in little packets that he called quanta, the energy of each quantum being proportional to the frequency of the radiation. The revolution gathered strength in 1905 when Einstein described light as consisting of little quantum particles that maintain their separate existence not only when they are emitted but also while they are traveling from place to place. The next big step forward came in 1913 when Bohr described atoms as miniature solar systems, with electrons traveling in orbits around the nucleus like planets orbiting the sun, and the energies of the orbits

*Doubleday, 1966; Dover, 1985.

†Viking, 2007.

taking discrete values limited by quantum conditions. All through the years from 1900 to 1923, physicists were suffering from schizophrenia. They had been educated to believe that the laws of classical physics could explain everything, but the new quantum effects were confirmed by experiments and were obviously inconsistent with the classical laws.

The real quantum revolution started in 1923 when the French physicist Louis de Broglie proposed dropping the classical laws altogether and representing all material objects by waves. The Austrian Erwin Schrödinger found the wave equation that converted Broglie's vision of matter waves into a coherent theory. The years from 1925 to 1928 were the era of *Knabenphysik*, or "Boy Physics." The radical new ideas of quantum mechanics emerged in rapid succession from the brains of twenty-five-year-old boys, in particular from the brains of Heisenberg, Pauli, and Dirac, while the older generation, including Bohr and Einstein and Schrödinger and Ehrenfest, struggled to keep up with them.

By 1932, when the *Faust* spoof was performed, the revolution was over. Quantum mechanics was firmly established. Dirac had announced the end of the revolution in 1929 with his customary clarity: "The underlying physical laws necessary for the mathematical theory of a large part of physics and the whole of chemistry are thus completely known." One of the main themes of Delbrück's script was the fact that the boy geniuses who invented quantum mechanics in 1925 were in 1932 already growing old. At the end of the play, Dirac makes another clear statement:

> *... Old age is a cold fever*
> *That every physicist suffers with!*
> *When one is past thirty,*
> *He is as good as dead!*

Heisenberg adds a fiercer tone to Dirac's lament: "It would be best to give them an early death." Finally, Pauli, who in real life was never at a loss for a word, ends the play with a sad confession: "Pauli has here nothing more to say!"

The play ends, and Pauli's reign as Mephistopheles is over. Delbrück is proclaiming to his twenty-five-year-old friends in the audience that the thirty-year-old wunderkinder in the front row are fading, and it is now time for the twenty-five-year-olds to take over the leadership of the revolution. It becomes clear at the end that Delbrück's sharpest satire is not directed against Ehrenfest but against the thirty-year-old geniuses who have too soon become elder statesmen.

In his subtitle, Segrè calls the Copenhagen performance "A Struggle for the Soul of Physics." The subtitle is not an accurate description, either of the performance or of the book. The performance was hardly concerned with physics at all. It was concerned with a remarkable group of human beings who had worked together for many years and achieved an amazing success. The performance celebrated their success by turning it into a comedy, using the pompous language of Goethe to make fun of their personal idiosyncrasies. It is a portrait of the group, seen in the distorting mirror of Delbrück's wit. Segrè's book is concerned with physics, but not with a struggle for the soul. It gives a lively account of the quantum revolution, interspersed with extracts from Goethe and Delbrück to add personal color to the narrative. Only at the end is there a brief passage describing the struggles for the soul of physics that began ten years later and had little to do with the quantum revolution.

There were two separate struggles for the soul of physics. One began when physics was applied on a gigantic scale to the production of nuclear weapons in World War II. Another began when physics after the war was increasingly dominated by big particle accelerators with large teams of scientists and engineers to operate them. Neither of these struggles was foreseen by Delbrück or by anyone else in 1932. The chief worry of the physicists in 1932 was the danger that they might run out of ideas. They did not worry about being taken over by the military or by heavy industry. They did not worry about losing their souls. Delbrück saw *Faust* as a convenient source of literary quotations, not as a moral dilemma for physicists. The idea that physicists working on nuclear energy were making a Faustian bargain with the devil came later, after the discovery of fission in 1938. The earliest such bargains were made by Heisenberg in Berlin in 1939, and by Bohr and many others at Los Alamos in 1943. Neither Heisenberg nor Bohr ever expressed remorse for the bargains that they made. Both of them remained firm believers in the promise of nuclear energy as a boon to all mankind.

The main question that the book raises is whether the quantum revolution of the 1920s was a unique event in the history of science, or whether it may some day recur. The generation of physicists who lived through it were mostly convinced that they would live to see it repeated. The experience of living through the crisis affected them so deeply that they could not easily return to less adventurous ways of thinking. They saw that the quantum revolution was incomplete and left many important mysteries unresolved. They could not give up the hope that they could solve the remaining mysteries with a second explosion of new ideas.

Many of the leaders of the first revolution, like Einstein, spent the rest of their lives pursuing various radical ideas that led nowhere. Each of them imagined that his own personal vision would be the key that would open the door to the second revolution. Their radical ideas were all different, but had in common the lack of any experimental support. The first revolution had been guided and tested by numerous experiments in atomic physics. The later radical ideas were not only untested but untestable. They did not make predictions that were precise enough to be proved right or wrong. Einstein had his unified field theories, bringing together the equations of classical electromagnetism and gravitation. Heisenberg had a nonlinear quantum field theory, which he promoted with great publicity and little success. Even Dirac, who was generally the most levelheaded of the group, spent some years pursuing a crazy version of quantum mechanics in which probabilities were allowed to be greater than one or less than zero. All these efforts failed, and the second revolution did not happen.

The only one of the older generation of revolutionaries who did not succumb to fantasies of a second revolution was Bohr. He remained until the end of his life actively engaged in supporting and encouraging successive generations of young scientists. He did not, like Einstein, retreat into an ivory tower to pursue his own ideas in isolation. When I was a young scientist at the Institute for Advanced Study in Princeton, I had occasion to observe at first hand the contrasting styles of Bohr and Einstein. That was in the early 1950s, when Bohr came to the institute among a crowd of younger visitors. He attended our seminars and took part in our arguments. He was interested in everything that we were doing. He enjoyed watching the science of particle physics unfold with frequent discoveries of unexpected particles and interactions. He was confident that the quantum revolution of the 1920s had provided a firm basis for understanding the new discoveries. He did not see any need for a second revolution.

At the same time, Einstein was working by himself in a nearby office, trying out one set of unified field equations after another. He never came to our seminars and never showed any interest in our activities. For us and for Bohr, the central problem of physics was to understand and explain the new particles. For Einstein, the new particles were uninteresting. He did not allow them to distract him from his chosen path. They never appeared in any of his equations.

Einstein and Bohr continued to move along divergent trajectories. Einstein was driven by a divine discontent that led him to reject the first quantum revolution and strive to create a second revolution out of pure thought. Bohr was driven by pride in the successes of the first revolution, which led him to continue exploring the details of nuclear and particle physics and enjoying the friendship of new generations of young scientists who came to work with him. The younger generations were faced with a choice between two alternatives. Should they follow Bohr and be content with a lifetime of solid but unrevolutionary research in the established fields of physics? Or should they follow Einstein and spend their lives in a lonely attempt to start a new revolution without any experimental guidance? They were caught in a trap, forced to choose between two paths, one leading to conservative mediocrity and the other to radical irrelevance. Physics was a trap, because the first revolution had already happened, and the only way to attempt another revolution was to jump into a hyperspace of pure speculation.

Delbrück and Gamow, the two progenitors of the Copenhagen *Faust*, found an escape from the trap. The way of escape was to move out from physics into other fields where revolutions had not yet happened. In other fields, revolutions were overdue, and it was still possible to start one without losing touch with reality. It was possible to be radical without being irrelevant. Gamow jumped from physics into cosmology, Delbrück from physics into biology, and both of them started revolutions.

Gamow revolutionized cosmology with his theory that the expansion of the universe started with a hot big bang. He proposed that the early universe was an explosively hot and dense mixture of particles and radiation, and his theory was testable because a relic of the early high-temperature radiation could still be detected. He predicted that a uniform sea of microwave radiation should still be pervading the universe today, with wavelengths increased and temperatures diminished by a factor of a thousand since the time when the universe was an opaque primeval fireball. According to his theory, this microwave background radiation should be barely intense enough to be detectable with sensitive radio telescopes. Three years before Gamow died, the cosmic microwave radiation was discovered by Arno Penzias and Robert Wilson, and the hot big bang cosmology became generally accepted as a true picture of the early universe.

Delbrück started a revolution in biology by choosing the bacteriophage, a simple type of virus that infects bacteria, as the object to be studied in detail. He observed that the revolution in physics had succeeded in large part because the hydrogen atom was chosen as the object of study. The hydrogen atom is the simplest kind of atom, consisting of a single proton and a single electron, and it has the simplest rules of behavior. Its behavior was simple enough to allow accurate comparisons of theory with experiment while the theory was being developed. So Delbrück chose the bacteriophage, or phage for short, as the hydrogen atom of biology. It was the simplest known form of life, and therefore the most likely to be intelligible.

To study the phage in detail was the most promising way to reach an understanding of life. First in Berlin, then at Vanderbilt University and the California Institute of Technology in the US, Delbrück organized a group of young scientists that he called the Phage Group. They studied phages with the tools of physics as well as the tools of biology. It turned out that the phage was well chosen as a key to some of the mysteries of life, but not to all of them. Life has two main functions: metabolism and replication. Metabolism is the complicated network of chemical processes that enable a living cell to maintain its integrity in a variable environment. Replication is the much simpler process of chemical copying that enables a parent cell to duplicate itself and produce two daughters. The phage is the simplest kind of organism because it has only replication and no metabolism. It is a pure parasite, replicating itself within a bacterium and borrowing the metabolic apparatus of the bacterium to perform its missing metabolic functions. The phage allowed Delbrück to elucidate the basic rules of replication without the complications associated with metabolism. The phage was in fact, as he had surmised at the beginning, a good substitute for the hydrogen atom.

Bohr's understanding of quantum mechanics was based on a philosophical principle, which he called complementarity. Two descriptions of nature are said to be complementary when they are both true but cannot both be seen in the same experiment. In quantum mechanics, the wave picture and the particle picture of an electron or a light quantum are complementary. You see the wave picture when you do an experiment with electrons or light reflected from a diffraction grating and observe the diffracted waves. You see the particle picture when you detect the electrons or light quanta in an electronic counter and count them one at a time. Complementarity in quantum mechanics is an established fact. But in 1932 Bohr proposed to extend the idea of complementarity to biology, suggesting that the description of a living creature as an organism and the description of it as a collection of molecules are also complementary. In this context, complementarity would mean that any attempt to observe and localize precisely every molecule in a living creature would result in the death of the organism. The holistic view of a creature as a living organism and the reductionist view of it as a collection of molecules would be both correct but mutually exclusive. Bohr believed strongly in this application of complementarity to the understanding of life. Delbrück believed in it too when he decided to become a biologist.

It is one of the ironies of history that Delbrück chose to study the phage, which may be the only organism simple enough to be described without invoking complementarity. The life of the phage is pure replication without metabolism. Replication is a chemical process that was completely explained by the double-helix structure of the DNA molecule discovered by Francis Crick and James Watson in 1953. When Crick and Watson discovered the double helix, they loudly claimed to have discovered the basic secret of life. The discovery came as a disappointment to Delbrück. It seemed to make complementarity unnecessary. Delbrück said it was as if the behavior of the hydrogen atom had been completely explained without requiring quantum mechanics. He recognized the importance of the discovery, but sadly concluded that it proved Bohr wrong. Life was, after all, simply and cheaply explained by looking in detail at a molecular model. Deep ideas of complementarity had no place in biology.

Segrè agrees with this judgment. He says dogmatically, "Bohr's conjecture was provocative, as it was meant to be, but in the end it turned out to be wrong. DNA and RNA are the answer to life, not complementarity." In the middle years of the twentieth century, this was the verdict of the majority of scientists. The historic debate over complementarity between Bohr and Einstein was over. Bohr had won in physics. Einstein had won in biology.

Now, fifty years later, Segrè's opinion is widely held by physicists, less widely by biologists. I disagree with it profoundly. In my opinion, the double helix is much too simple to be the secret of life. If DNA had been the secret of life, we should have been able to cure cancer long ago. The double helix explains replication but it does not explain metabolism. Delbrück chose to study the phage because it embodies replication without metabolism, and Crick and Watson chose to study DNA for the same reason. Replication is clean while metabolism is messy. By excluding messiness, they excluded the essence of life. The genomes of human and other creatures have now been completely mapped and the processes of replication have been thoroughly explored, but the mysteries of metabolism still remain mysteries.

The phage is still the only living creature whose behavior is simple enough to be completely understood and predicted. To understand other kinds of creatures, from fruit flies to humans, we need also a deep understanding of metabolism. The understanding of metabolism will perhaps be the theme of the next revolution in biology. I have already discussed a seminal paper by the biologist Carl Woese with the title "A New Biology for a New Century," pointing the way toward the next revolution.* Woese's new biology is based on the idea that a living creature is a dynamic pattern of organization in the stream of chemical materials and energy that passes through it. Patterns of organization are constantly forming and reforming themselves. If we try to observe and localize every molecule as it passes through an organism, we are likely to destroy the patterns that constitute metabolic life. In Woese's picture of life, complementarity plays a central role, just as Bohr said it should.

At the same time, while Woese and others are debating the future of biology, the great debate over the future of physics continues. It is still a debate over the same questions that caused the disagreement between Bohr and Einstein. Does the quantum theory of the 1920s, together with the standard model of particles and interactions that grew out of it, give us a solid foundation for understanding nature? Or do we need another revolution to reach a deeper understanding?

Theoretical physicists are now divided into two main factions. Those who look forward to another revolution mostly believe that it will grow out of a grand mathematical scheme known as string theory. Those who are content with the outcome of the old revolution are mostly studying more mundane subjects such as high-temperature superconductors and quantum computers. String theory may be considered to be the counterattack of those who lost the debate over

*See chapter 1, "Our Biotech Future."

complementarity in physics in Copenhagen in 1932. It is the revenge of the heirs of Einstein against the heirs of Bohr. The new discipline of systems biology, describing living creatures as emergent dynamic organizations rather than as collections of molecules, is the counter-attack of those who lost the debate over complementarity in biology in 1953. It is the revenge of the heirs of Bohr against the heirs of Einstein.

6

THE QUESTION OF GLOBAL WARMING

I BEGIN THIS review with a prologue, describing the measurements that transformed global warming from a vague theoretical speculation into a precise observational science.

There is a famous graph showing the fraction of carbon dioxide in the atmosphere as it varies month by month and year by year (see below). It gives us our firmest and most accurate evidence of the effects

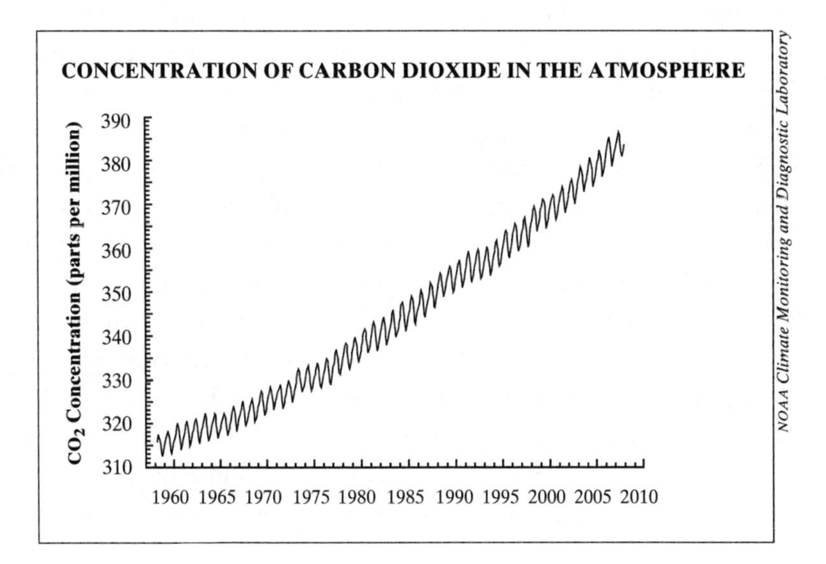

of human activities on our global environment. The graph is generally known as the Keeling graph because it summarizes the lifework of Charles David Keeling, a professor at the Scripps Institution of Oceanography in La Jolla, California. Keeling measured the carbon dioxide abundance in the atmosphere for forty-seven years, from 1958 until his death in 2005. He designed and built the instruments that made accurate measurements possible. He began making his measurements near the summit of the dormant volcano Mauna Loa on the big island of Hawaii.

He chose this place for his observatory because the ambient air is far from any continent and is uncontaminated by local human activities or vegetation. The measurements have continued after Keeling's death, and show an unbroken record of rising carbon dioxide abundance extending over fifty years. The graph has two obvious and conspicuous features. First, a steady increase of carbon dioxide with time, beginning at 315 parts per million in 1958 and reaching 385 parts per million in 2008. Second, a regular wiggle showing a yearly cycle of growth and decline of carbon dioxide levels. The maximum happens each year in the Northern Hemisphere spring, the minimum in the Northern Hemisphere fall. The difference between maximum and minimum each year is about six parts per million.

Keeling was a meticulous observer. The accuracy of his measurements has never been challenged, and many other observers have confirmed his results. In the 1970s he extended his observations from Mauna Loa, at latitude 20 north, to eight other stations at various latitudes, from the South Pole at latitude 90 south to Point Barrow on the Arctic coast of Alaska at latitude 71 north. At every latitude there is the same steady growth of carbon dioxide levels, but the size of the annual wiggle varies strongly with latitude. The wiggle is largest at Point Barrow where the difference between maximum and minimum is about fifteen parts per million. At Kerguelen, a Pacific island at latitude 29 south, the wiggle vanishes. At the South Pole the difference between maximum and minimum is about two parts per million, with the maximum in the Southern Hemisphere spring.

The only plausible explanation of the annual wiggle and its variation with latitude is that it is due to the seasonal growth and decay of annual vegetation, especially deciduous forests, in temperate latitudes north and south. The asymmetry of the wiggle between north and south is caused by the fact that the Northern Hemisphere has most of the land area and most of the deciduous forests. The wiggle is giving us a direct measurement of the quantity of carbon that is absorbed from the atmosphere each summer north and south by growing vegetation, and returned each winter to the atmosphere by dying and decaying vegetation.

The quantity is large, as we see directly from the Point Barrow measurements. The wiggle at Point Barrow shows that the net growth of vegetation in the Northern Hemisphere summer absorbs about 4 percent of the total carbon dioxide in the high-latitude atmosphere each year. The total absorption must be larger than the net growth, because the vegetation continues to respire during the summer, and the net growth is equal to total absorption minus respiration. The tropical forests at low latitudes are also absorbing and respiring a large quantity of carbon dioxide, which does not vary much with the season and does not contribute much to the annual wiggle.

When we put together the evidence from the wiggles and the distribution of vegetation over the earth, it turns out that about 8 percent of the carbon dioxide in the atmosphere is absorbed by vegetation and returned to the atmosphere every year. This means that the average lifetime of a molecule of carbon dioxide in the atmosphere, before it is captured by vegetation and afterward released, is about twelve years. This fact, that the exchange of carbon between atmosphere and vegetation is rapid, is of fundamental importance to the long-range future of global warming, as will become clear in what follows. Neither of the books under review mentions it.

William Nordhaus is a professional economist, and his book *A Question of Balance: Weighing the Options on Global Warming Policies* describes the global-warming problem as an economist sees it.* He is not concerned with the science of global warming or with the detailed estimation of the damage that it may do. He assumes that the science and the damage are specified, and he compares the effectiveness of various policies for the allocation of economic resources in response. His conclusions are largely independent of scientific details. He calculates aggregated expenditures and costs and gains. Everything is calculated by running a single computer model that he calls DICE, an acronym for Dynamic Integrated model of Climate and the Economy.

Each run of DICE takes as input a particular policy for allocating expenditures year by year. The allocated resources are spent on subsidizing costly technologies—for example, deep underground sequestration of carbon dioxide produced in power stations—that reduce emissions of carbon dioxide, or placing a tax on activities that produce carbon emissions. The climate model part of DICE calculates the effect of the reduced emissions in reducing damage. The output of DICE then tells us the resulting gains and losses of the world economy year by year. Each run begins at 2005 and ends either at 2105 or 2205, giving a picture of the effects of a particular policy over the next one or two hundred years.

The practical unit of economic resources is a trillion inflation-adjusted dollars. An inflation-adjusted dollar means a sum of money,

*Yale University Press, 2008.

at any future time, with the same purchasing power as a real dollar in 2005. In the following discussion, the word "dollar" will always mean an inflation-adjusted dollar, with a purchasing power that does not vary with time. The difference in outcome between one policy and another is typically several trillion dollars, comparable with the cost of the war in Iraq. This is a game played for high stakes.

Nordhaus's book is not for the casual reader. It is full of graphs and tables of numbers, with an occasional equation to show how the numbers are related. The graphs and tables show how the world economy reacts to the various policy options. To understand these graphs and tables, readers should be familiar with financial statements and compound interest, but they do not need to be experts in economic theory. Anyone who knows enough mathematics to balance a checkbook or complete an income tax return should be able to understand the numbers.

For the benefit of those who are mathematically illiterate or uninterested in numerical details, Nordhaus has put a nonmathematical chapter at the beginning with the title "Summary for the Concerned Citizen." This first chapter contains an admirably clear summary of his results and their practical consequences, digested so as to be read by busy politicians and ordinary people who may vote the politicians into office. He believes that the most important concern of any policy that aims to address climate change should be how to set the most efficient "carbon price," which he defines as "the market price or penalty that would be paid by those who use fossil fuels and thereby generate CO_2 emissions." He writes:

> Whether someone is serious about tackling the global-warming problem can be readily gauged by listening to what he or she says about the carbon price. Suppose you hear a public figure who speaks eloquently of the perils of global warming and proposes

> that the nation should move urgently to slow climate change. Suppose that person proposes regulating the fuel efficiency of cars, or requiring high-efficiency lightbulbs, or subsidizing ethanol, or providing research support for solar power—but nowhere does the proposal raise the price of carbon. You should conclude that the proposal is not really serious and does not recognize the central economic message about how to slow climate change. To a first approximation, raising the price of carbon is a necessary and sufficient step for tackling global warming. The rest is at best rhetoric and may actually be harmful in inducing economic inefficiencies.

If this chapter were widely read, the public understanding of global warming and possible responses to it would be greatly improved.

Nordhaus examines five kinds of global-warming policy, with many runs of DICE for each kind. The first kind is "business as usual," with no restriction of carbon dioxide emissions—in which case, he estimates damages to the environment amounting to some $23 trillion in current dollars by the year 2100. The second kind is the "optimal policy," judged by Nordhaus to be the most cost-effective, with a worldwide tax on carbon emissions adjusted each year to give the maximum aggregate economic gain as calculated by DICE. The third kind is the Kyoto Protocol, in operation since 2005 with 175 participating countries, imposing fixed limits to the emissions of economically developed countries only. Nordhaus tests various versions of the Kyoto Protocol, with or without the participation of the United States.

The fourth kind of policy is labeled "ambitious" proposals, with two versions that Nordhaus calls "Stern" and "Gore." "Stern" is the policy advocated by Sir Nicholas Stern in the *Stern Review*, an economic analysis of global-warming policy sponsored by the British government.* "Stern" imposes draconian limits on emissions, similar to the Kyoto limits but much stronger. "Gore" is a policy advocated by Al Gore, with emissions reduced drastically but gradually, the reductions reaching 90 percent of current levels before the year 2050. The fifth and last kind is called "low-cost backstop," a policy based on a hypothetical low-cost technology for removing carbon dioxide from the atmosphere, or for producing energy without carbon dioxide emission, assuming that such a technology will become available at some specified future date. According to Nordhaus, this technology might include "low-cost solar power, geothermal energy, some nonintrusive climatic engineering, or genetically engineered carbon-eating trees."

Since each policy put through DICE is allowed to run for one or two hundred years, its economic effectiveness must be measured by an aggregated sum of gains and losses over the whole duration of the run. The most crucial question facing the policymaker is then how to compare present-day gains and losses with gains and losses a hundred years in the future. That is why Nordhaus chose *A Question of Balance* for his title. If we can save M dollars of damage caused by climate change in the year 2110 by spending one dollar on reducing emissions in the year 2010, how large must M be to make the spending worthwhile? Or, as economists might put it, how much can future losses from climate change be diminished or "discounted" by money invested in reducing emissions now?

The conventional answer given by economists to this question is to say that M must be larger than the expected return in 2110 if the 2010 dollar were invested in the world economy for a hundred years at an average rate of compound interest. For example, the value of

*See Nicholas Stern, *The Economics of Climate Change: The Stern Review* (Cambridge University Press, 2007).

one dollar invested at an average interest rate of 4 percent for a period of one hundred years would be fifty-four dollars; this would be the future value of one dollar in one hundred years' time. Therefore, for every dollar spent now on a particular strategy to fight global warming, the investment must reduce the damage caused by warming by an amount that exceeds fifty-four dollars in one hundred years' time to accrue a positive economic benefit to society. If a strategy of a tax on carbon emissions results in a return of only forty-four dollars per dollar invested, the benefits of adopting the strategy will be outweighed by the costs of paying for it. But if the strategy produces a return of sixty-four dollars per dollar invested, the advantages are clear. The question then is how well different strategies of dealing with global warming succeed in producing long-term benefits that outweigh their present costs. The aggregation of gains and losses over time should be calculated with the remote future heavily discounted.

The choice of discount rate for the future is the most important decision for anyone making long-range plans. The discount rate is the assumed annual percentage loss in present value of a future dollar as it moves further into the future. The DICE program allows the discount rate to be chosen arbitrarily, but Nordhaus displays the results only for a discount rate of 4 percent. Here he is following the conventional wisdom of economists. Four percent is a conservative number, based on an average of past experience in good and bad times. Nordhaus is basing his judgment on the assumption that the next hundred years will bring to the world economy a mixture of stagnation and prosperity, with overall average growth continuing at the same rate that we have experienced during the twentieth century. Future costs are discounted because the future world will be richer and better able to afford them. Future benefits are discounted because they will be a diminishing fraction of future wealth.

When the future costs and benefits are discounted at a rate of 4 percent per year, the aggregated costs and benefits of a climate policy over the entire future are finite. The costs and benefits beyond a hundred years make little difference to the calculated aggregate. Nordhaus therefore takes the aggregate benefit-minus-cost over the entire future as a measure of the net value of the policy. He uses this single number, calculated with the DICE model of the world economy, as a figure of merit to compare one policy with another. To represent the value of a policy by a single number is a gross oversimplification of the real world, but it helps to concentrate our attention on the most important differences between policies.

Here are the net values of the various policies as calculated by the DICE model. The values are calculated as differences from the "business as usual" model, without any emission controls. A plus value means that the policy is better than "business as usual," with the reduction of damage due to climate change exceeding the cost of controls. A minus value means that the policy is worse than "business as usual," with costs exceeding the reduction of damage. The unit of value is $1 trillion, and the values are specified to the nearest trillion. The net value of the optimal program, a global carbon tax increasing gradually with time, is plus three—that is, a benefit of some $3 trillion. The Kyoto Protocol has a value of plus one with US participation, zero without US participation. The "Stern" policy has a value of minus fifteen, the "Gore" policy minus twenty-one, and "low-cost backstop" plus seventeen.

What do these numbers mean? One trillion dollars is a difficult unit to visualize. It is easier to think of it as $3,000 for every man, woman, and child in the US population. It is comparable to the annual gross domestic product of India or Brazil. A gain or loss of $1 trillion would be a noticeable but not overwhelming perturbation of the world economy. A gain or loss of $10 trillion would be a major perturbation with unpredictable consequences.

The main conclusion of the Nordhaus analysis is that the ambitious proposals, "Stern" and "Gore," are disastrously expensive; the "low-cost backstop" is enormously advantageous if it can be achieved; and the other policies, including "business as usual" and Kyoto, are only moderately worse than the optimal policy. The practical consequence for global-warming policy is that we should pursue the following objectives in order of priority: (1) Avoid the ambitious proposals. (2) Develop the science and technology for a low-cost backstop. (3) Negotiate an international treaty coming as close as possible to the optimal policy, in case the low-cost backstop fails. (4) Avoid an international treaty making the Kyoto Protocol policy permanent. These objectives are valid for economic reasons, independent of the scientific details of global warming.

There is a fundamental difference of philosophy between Nordhaus and Stern. Chapter 9 of Nordhaus's book explains the difference, and explains why Stern advocates a policy that Nordhaus considers disastrous. Stern rejects the idea of discounting future costs and benefits when they are compared with present costs and benefits. Nordhaus, following the normal practice of economists and business executives, considers discounting to be necessary for reaching any reasonable balance between present and future. In Stern's view, discounting is unethical because it discriminates between present and future generations. That is, Stern believes that discounting imposes excessive burdens on future generations. In Nordhaus's view, discounting is fair because a dollar saved by the present generation becomes fifty-four dollars to be spent by our descendants a hundred years later.

The practical consequence of the Stern policy would be to slow down the economic growth of China now in order to reduce damage from climate change a hundred years later. Several generations of

Chinese citizens would be impoverished to make their descendants only slightly richer. According to Nordhaus, the slowing down of growth would in the end be far more costly to China than the climatic damage. About the much-discussed possibility of catastrophic effects before the end of the century from rising sea levels, he says only that "climate change is unlikely to be catastrophic in the near term, but it has the potential for serious damages in the long run." The Chinese government firmly rejects the Stern philosophy, while the British government enthusiastically embraces it. The *Stern Review*, according to Nordhaus, "takes the lofty vantage point of the world social planner, perhaps stoking the dying embers of the British Empire."

The main deficiency of Nordhaus's book is that he does not discuss the details of the "low-cost backstop" that might provide a climate policy vastly more profitable than his optimum policy. He avoids this subject because he is an economist and not a scientist. He does not wish to question the pronouncements of the Intergovernmental Panel on Climate Change, a group of hundreds of scientists officially appointed by the United Nations to give scientific advice to governments. The IPCC considers the science of climate change to be settled and does not believe in low-cost backstops. Concerning the possible candidates for a low-cost backstop technology he mentions in the sentence I previously quoted—for example, "low-cost solar power"—Nordhaus has little to say. He writes that "no such technology presently exists, and we can only speculate on it." The "low-cost backstop" policy is displayed in his tables as an abstract possibility without any details. It is nowhere emphasized as a practical solution to the problem of climate change.

At this point I return to the Keeling graph, which demonstrates the strong coupling between atmosphere and plants. The wiggles in the graph show us that every carbon dioxide molecule in the atmosphere is incorporated in a plant within a time of the order of twelve years. Therefore, if we can control what the plants do with the carbon, the fate of the carbon in the atmosphere is in our hands. That is what Nordhaus meant when he mentioned "genetically engineered carbon-eating trees" as a low-cost backstop to global warming. The science and technology of genetic engineering are not yet ripe for large-scale use. We do not understand the language of the genome well enough to read and write it fluently. But the science is advancing rapidly, and the technology of reading and writing genomes is advancing even more rapidly. I consider it likely that we shall have "genetically engineered carbon-eating trees" within twenty years, and almost certainly within fifty years.

Carbon-eating trees could convert most of the carbon that they absorb from the atmosphere into some chemically stable form and bury it underground. Or they could convert the carbon into liquid fuels and other useful chemicals. Biotechnology is enormously powerful, capable of burying or transforming any molecule of carbon dioxide that comes into its grasp. Keeling's wiggles prove that a big fraction of the carbon dioxide in the atmosphere comes within the grasp of biotechnology every decade. If one quarter of the world's forests were replanted with carbon-eating varieties of the same species, the forests would be preserved as ecological resources and as habitats for wildlife, and the carbon dioxide in the atmosphere would be reduced by half in about fifty years.

It is likely that biotechnology will dominate our lives and our economic activities during the second half of the twenty-first century, just as computer technology dominated our lives and our economy during the second half of the twentieth. Biotechnology could be a great equalizer, spreading wealth over the world wherever there is land and air and water and sunlight. This has nothing to do with the misguided efforts that are now being made to reduce carbon emissions by growing corn and converting it into ethanol fuel. The ethanol program fails to reduce emissions and incidentally hurts poor people all over the world by raising the price of food. After we have mastered biotechnology, the rules of the climate game will be radically changed. In a world economy based on biotechnology, some low-cost and environmentally benign backstop to carbon emissions is likely to become a reality.

Global Warming: Looking Beyond Kyoto is the record of a conference held at the Yale Center for the Study of Globalization in 2005.* It is edited by Ernesto Zedillo, the head of the Yale Center, who served as the president of Mexico from 1994 to 2000 and was chairman of the conference. The book consists of an introduction by Zedillo and fourteen chapters contributed by speakers at the conference. Among the speakers was Nordhaus, contributing "Economic Analyses of the Kyoto Protocol: Is There Life After Kyoto?," a sharper criticism of the Kyoto Protocol than we find in his own book.

The Zedillo book covers a much wider range of topics and opinions than the Nordhaus book, and is addressed to a wider circle of readers. It includes "Is the Global Warming Alarm Founded on Fact?" by Richard Lindzen, a professor of atmospheric sciences at MIT, answering that question with a resounding no. Lindzen does not deny the existence of global warming, but considers the predictions of its harmful effects to be grossly exaggerated. He writes:

> Actual observations suggest that the sensitivity of the real climate is much less than that found in computer models whose sensitivity depends on processes that are clearly misrepresented.

*Yale Center for the Study of Globalization/Brookings Institution Press, 2008.

Answering Lindzen in the next chapter, "Anthropogenic Climate Change: Revisiting the Facts," is Stefan Rahmstorf, a professor of physics of the oceans at Potsdam University in Germany. Rahmstorf sums up his opinion of Lindzen's arguments in one sentence: "All this seems completely out of touch with the world of climate science as I know it and, to be frank, simply ludicrous." These two chapters give the reader a sad picture of climate science. Rahmstorf represents the majority of scientists who believe fervently that global warming is a grave danger. Lindzen represents the small minority who are skeptical. Their conversation is a dialogue of the deaf. The majority responds to the minority with open contempt.

In the history of science it has often happened that the majority was wrong and refused to listen to a minority that later turned out to be right. It may—or may not—be that the present is such a time. The great virtue of Nordhaus's economic analysis is that it remains valid whether the majority view is right or wrong. Nordhaus's optimum policy takes both possibilities into account. Zedillo in his introduction summarizes the arguments of each contributor in turn. He maintains the neutrality appropriate to a conference chairman and gives equal space to Lindzen and to Rahmstorf. He betrays his own opinion only in a single sentence with a short parenthesis: "Climate change may not be the world's most pressing problem (as I am convinced it is not), but it could still prove to be the most complex challenge the world has ever faced."

The last five chapters of the Zedillo book are by writers from five of the countries most concerned with the politics of global warming: Russia, Britain, Canada, India, and China. Each of the five authors has been responsible for giving technical advice to a government, and each of them gives us a statement of that government's policy. Howard Dalton, spokesman for the British government, is the most dogmatic. His final paragraph begins:

> It is the firm view of the United Kingdom that climate change constitutes a major threat to the environment and human society, that urgent action is needed now across the world to avert that threat, and that the developed world needs to show leadership in tackling climate change.

The United Kingdom has made up its mind and takes the view that any individuals who disagree with government policy should be ignored. This dogmatic tone is also adopted by the Royal Society, the British equivalent of the US National Academy of Sciences. The Royal Society recently published a pamphlet addressed to the general public with the title "Climate Change Controversies: A Simple Guide." The pamphlet says:

> This is not intended to provide exhaustive answers to every contentious argument that has been put forward by those who seek to distort and undermine the science of climate change and deny the seriousness of the potential consequences of global warming.

In other words, if you disagree with the majority opinion about global warming, you are an enemy of science. The authors of the pamphlet appear to have forgotten the ancient motto of the Royal Society, *Nullius in Verba*, which means, "Nobody's word is final."

All the books that I have seen about the science and economics of global warming, including the two books under review, miss the main point. The main point is religious rather than scientific. There is a worldwide secular religion that we may call environmentalism, holding that we are stewards of the earth, that despoiling the planet with waste products of our luxurious living is a sin, and that the path of righteousness is to live as frugally as possible. The ethics of environmentalism are being taught to children in kindergartens, schools, and colleges all over the world.

Environmentalism has replaced socialism as the leading secular religion. And the ethics of environmentalism are fundamentally sound. Scientists and economists can agree with Buddhist monks and Christian activists that ruthless destruction of natural habitats is evil and careful preservation of birds and butterflies is good. The world-wide community of environmentalists—most of whom are not scientists—holds the moral high ground and is guiding human societies toward a hopeful future. Environmentalism, as a religion of hope and respect for nature, is here to stay. This is a religion that we can all share, whether or not we believe that global warming is harmful.

Unfortunately, some members of the environmental movement have also adopted as an article of faith the belief that global warming is the greatest threat to the ecology of our planet. That is one reason why the arguments about global warming have become bitter and passionate. Much of the public has come to believe that anyone who is skeptical about the dangers of global warming is an enemy of the environment. The skeptics now have the difficult task of convincing the public that the opposite is true. Many of the skeptics are passionate environmentalists. They are horrified to see the obsession with global warming distracting public attention from what they see as more serious and more immediate dangers to the planet, including problems of nuclear weaponry, environmental degradation, and social injustice. Whether they turn out to be right or wrong, their arguments on these issues deserve to be heard.

Note added in 2014: This review stimulated a flow of letters and responses, too voluminous to be summarized here. The most useful exchange was with Lord May, who is one of the most influential scientists in the United Kingdom, having been the chief scientific adviser to the British government (1995–2000) and president of the Royal Society (2000–2005). Here are extracts from his letter and my reply:

> *Lord May:* The essay begins with a characteristically clear and elegant exposition of the annual uptake of carbon dioxide by plants, and the subsequent reemission from respiration or decay. This leads Dyson to the conclusion that "the average lifetime of a molecule of carbon dioxide in the atmosphere...is about twelve years." Dyson correctly emphasizes that such a timescale is fundamental to discussions of global warming. Unfortunately, however, estimates of the characteristic "residence time" of a molecule of carbon dioxide in the atmosphere involve a complicated mélange of factors, leading to the conclusion that although almost half of newly added carbon dioxide molecules remain for only a decade or two, roughly a third stay for a century or more, and fully one fifth for a millennium.... This is why the residence time of such molecules is generally characterized as a century.
>
> This difference between twelve versus one hundred years is no minor pedantry. A major argument for the need for urgent action now, even though truly major consequences of global warming may lie some decades ahead, is that the carbon dioxide molecules we are putting into the atmosphere today are going to hang around, continuing to thicken the greenhouse gas blanket, for a long time.
>
> *Dyson:* Lord May and I have several differences of opinion, which remain friendly. But one of our disagreements is a matter

of arithmetic and not a matter of opinion. He says that the residence time of a molecule of carbon dioxide in the atmosphere is about a century, and I say it is about twelve years.

This discrepancy is easy to resolve. We are talking about different meanings of residence time. I am talking about residence without replacement. My residence time is the time that an average carbon dioxide molecule stays in the atmosphere before being absorbed by a plant. He is talking about residence with replacement. His residence time is the average time that a carbon dioxide molecule and its replacements stay in the atmosphere when, as usually happens, a molecule that is absorbed is replaced by another molecule emitted from another plant.... In my review I was discussing the use of carbon-eating plants to sequester carbon dioxide from the atmosphere.... Since we are discussing the effect of carbon-eating plants, my use of the short residence time without replacement is correct, and his use of the long residence time with replacement in that situation is wrong.

7

STRUGGLE FOR THE ISLANDS

THE MOST DRAMATIC moment of our trip to the Galápagos Islands in May 2008 was on the last day. My wife and I were leaning over the railing on the deck of the tourist boat *Integrity*, watching an orca whale. The orca swam close to the boat, almost directly underneath us. Then, just ahead of the orca, a large sea turtle appeared. This was not one of the giant tortoises for which the islands are famous but an equally massive marine turtle. The females of the species come to the islands to lay their eggs under the sand on the beaches.

My wife had met this turtle earlier in the day, when she was swimming in the ocean with a snorkel. Only a second after we saw the turtle from the boat, the orca snapped, biting through the turtle shell as if it were a pie crust. Immediately the sea turned red and fifty frigate birds appeared from nowhere to pick up the larger remaining scraps of flesh. After the frigate birds were done, flocks of smaller birds came to pick up the smaller scraps. The red sea rapidly faded. In less than a minute it was all over. It was like a scene from *National Geographic* on television, but real.

Perhaps we were partly responsible for the turtle's death, since the turtle and the orca were both attracted to the boat. If we had not come to disturb the normal rhythm of her life, the turtle might now be out of harm's way, mother to a new batch of hatchlings. But in the ordinary course of nature, without boats and tourists, such a death is not unusual. We had seen nature doing her daily work, holding the balance impartially between predator and prey. Only in our eyes is nature beautiful and cruel.

Galápagos: The Islands That Changed the World is a combination of four books in one.* It is first a picture book, second a guidebook, third a history book, and fourth a political manifesto. I will describe the four components in turn and then reflect upon their message. The picture book is a gallery of magnificent photographs of the islands and their nonhuman inhabitants, taken over many years by Tui De Roy and others. De Roy is a professional photographer who arrived on the islands with her family at the age of two and spent much of her life there. About fifty of the pictures are hers, including two portraits: of Darwin's finches and of blue-footed boobies.

Darwin's finches are inconspicuous little birds that Darwin observed when he visited the islands in 1835. They later provided crucial evidence for his theory of the origin of species. Blue-footed boobies are big seabirds that walk around the islands on bright blue duck-feet. These two images exemplify the clash of cultures that compete in historic places around the world: the culture of preservation and the culture of exploitation. Scholars and scientists try to preserve historic sites, while local entrepreneurs try to exploit them. Darwin's finches are the chief attraction of the Galápagos for professional biologists and historians of science. Blue-footed boobies are the chief attraction for sellers of souvenirs in tourist shops.

Other photographs in the book were taken by Daniel Fitter, who was our guide on the island of Santa Cruz. All visitors to the national park must be accompanied by a licensed guide. He walked with us

---

*Paul Stewart et al. (Yale University Press, 2007).

into the farmland to find giant tortoises, who choose to live comfortably in the small irrigated area open to human settlement rather than in the more austere environment of the national park. One of Fitter's photographs shows the small island Daphne Major, an uninhabited volcanic crater, silhouetted against a threatening sky. Daphne Major is famous as the worksite of Peter and Rosemary Grant, who camped there for several months every year for twenty years, laboriously studying the birds and incidentally raising two daughters.

The island is small and the birds are tame enough so that the Grants could catch and label every finch that lived there and record its individual life history, from hatching and mating to parenting and mortality. They assigned each finch to one of the thirteen endemic species by measuring the size and shape of its beak. "Endemic" means a species that breeds in the islands and nowhere else. They discovered an astonishing fact that Darwin missed: evolution by natural selection sometimes moves fast. Darwin imagined that evolution must be slower than any possible human observation, requiring thousands or millions of years to form new species. The Grants observed hybridization and segregation of species happening within a few years, fast enough to be seen and accurately measured by humans.

The reason why evolution in the Galápagos is fast is that climate and vegetation change abruptly from year to year, and natural selection is brutal. Wet and dry years unpredictably produce lush and sparse vegetation. In lush years, there are plenty of small, soft seeds, and birds with smaller beaks and quicker reproduction have an advantage. In drought years, soft seeds are scarce, and birds with larger beaks specialized to deal with unusually large, tough seeds have an advantage. Selection is fast because populations of birds with the wrong kind of beak to split seeds that happen to be abundant may be wiped out in a single season.

The life and work of the Grants is described in an excellent book,

The Beak of the Finch, by Jonathan Weiner (1994), with hand-drawn illustrations by Thalia Grant and Charles Darwin.* Both of them are gifted artists. Thalia is one of the two Grant daughters who were raised on Daphne Major. Before we came to the Galápagos in 2008, we met the Grants by chance at a lunch party in Princeton. They told us that May was the best time to visit, at the transition between the wet and dry seasons. It was clearly understood that we were coming as tourists, not as scientists, and that we were not coming to Daphne Major.

The second component of *Galápagos* is an illustrated handbook for tourists, describing the best places and times to go walking or swimming or diving, and identifying the species of birds and reptiles and fish that the visitor will find. The handbook occupies the last fifty pages of the book. It is not intended for experts. It provides only brief descriptions of the thirteen species of Darwin's finches, with a picture of only one of them. It describes three species of shark but only illustrates one. An expert birdwatcher or scuba diver would need a more technical and specialized handbook. This one is aimed at the average tourist who is not interested in fine distinctions between closely related species of birds and fish. For the average tourist, a visit to the Galápagos is a unique experience because of the overwhelming abundance of the populations of a few species. The number of species on the islands is not large, but a small number of them have unusually dense populations. The populations of the dominant species seem even larger than they are, because the wild creatures are unafraid and do not move away when humans walk among them.

We were lucky to arrive on the island of Española during the breeding season of the albatrosses. The handbook tells us that these birds

*See also Peter Grant and Rosemary Grant, *How and Why Species Multiply: The Radiation of Darwin's Finches* (Princeton University Press, 2007).

weigh ten pounds and have a wingspan of eight feet. They live for thirty or forty years and generally mate for life. They are magnificent flyers but have difficulty with taking off and landing. Almost the entire world population of this species comes to Española to breed. We walked for miles over the island, placing our feet carefully so as not to step on the birds or their eggs. The ground was covered with majestic birds, each pair guarding a single egg, father and mother taking equal shares of the egg-sitting. In the distance over the ocean, the sky was thick with absent parents taking turns hunting for fish. The island shows what happens to a population when food is abundant and predators are lacking. These birds evolved to fly long distances over the ocean. When local fish are scarce they can find plenty of fish farther away. Albatrosses, in the region of Española where we walked, seemed to cover the ground as densely as humans in Manhattan.

Another striking photograph by De Roy shows a flightless cormorant spreading its wings to dry after a swim. The cormorant evolved in the opposite direction from the albatross, reducing the size of its wings until it lost the ability to fly. Cormorants and albatrosses coexist peacefully because they occupy separate ecological niches, the cormorant fishing close to shore and the albatross farther out, the cormorant supreme as a swimmer and the albatross as a flyer. The cormorants on Española are far less abundant than the albatrosses. The population of cormorants is limited by the population of fish within diving range of the shore. We did not need to step over cormorants as we walked, since they occupy only the high rocks overlooking the ocean.

The third and fourth components of *Galápagos*, the history book and the political manifesto, together make up the rest of the text, written by six authors. Each author has a single chapter, except for the chief author, Paul Stewart, who has four. The authors of single chapters are Patrick Morris on the geological history of the islands,

Andrew Murray on the history of Darwin's visit and his slow understanding of the creatures that he found there, Joe Stevens on the diverse ecologies of the coasts, Richard Wollocombe on the oceanic environment, and Godfrey Merlen on the successes and failures of conservation. Stewart wrote a prologue chapter, a chapter on human discovery and settlement, a chapter on the flora and fauna, and a concluding chapter with the title "Galápagos—World's End." The authors lived and worked together on the islands, producing the BBC television series *Galápagos*, of which this book is a summary. Roughly speaking, the chapters by Morris, Murray, Stevens, and Wollocombe are the history book, and the chapters by Merlen and Stewart are the political manifesto.

The centerpiece of the history book is the Darwin chapter. When Darwin arrived on the *Beagle*, he was mainly interested in the geology of the islands rather than the biology. The islands are the tops of a group of volcanoes. They are spectacularly young, some with craters still hot from recent eruptions, others with twisted ropes of newly solidified lava stretched along the shore. After he arrived, not yet looking for biological treasures, Darwin encountered a dense concentration of black iguanas sunbathing on the shore and grazing on seaweed. He wrote in his notebook without enthusiasm:

> The black Lava rocks on the beach are frequented by large (2–3 ft.) most disgusting clumsy Lizards. They are as black as the porous rocks over which they crawl.

But he soon realized that the animals were more interesting than the rocks. He knew enough zoology to understand that a unique population of creatures was living on these isolated islands. They must have appeared on the islands recently, after the islands emerged from the sea.

After the iguanas, Darwin met the tortoises, and after the tortoises, the birds. He went after the birds with his shotgun, collecting their skins, which were small enough to be conveniently shipped home to England. He observed that there were three closely similar species of mockingbirds, but on each island there was only one species. This gave him the idea that a single species had arrived on the islands and split into three after settling on different islands. He wrote in his notebook, "Such facts would undermine the stability of Species." He did not do so well with the finches. He collected large numbers of them but labeled members of various finch species as blackbirds and wrens and woodpeckers. After he arrived in London a year later, he showed the specimens to John Gould, an expert on birds. It was Gould who discovered that the birds now known as Darwin's finches are a closely related group of thirteen species. Darwin then saw that the finches provided stronger evidence than the mockingbirds of a single ancestral species splitting into many daughter species specialized to various environments and various ways of life.

The fourth component of the book is the political manifesto. The book, like the television series, has a political agenda, most visible in the chapters written by Stewart. The agenda is a view of the world that may be called "Doom-and-Gloom Environmentalism," or perhaps more accurately "Black-and-White Environmentalism." The world is seen as sharply divided into black and white, with no shades of gray. White is wilderness, the natural environment as it was before humans invaded and ruined it. Small patches of white survive in protected areas such as the Galápagos and parts of the Amazonian jungle. Black is cities and roads and shopping malls and parking lots, the areas where nature has been expelled and human constructions dominate. Since world population and industry are growing, the black areas are increasing and the white areas decreasing. The planet is inexorably turning black, and our only hope of improvement is a radical change in our way of living. Since any radical change is unlikely, the appropriate response of an enlightened person is gloom and doom.

According to the black-and-white environmentalist view, the only possible futures for the islands are to be totally white, a natural wilderness preserved from human disturbance, or to be totally black, a community of human predators with no respect for nature. They can only be preserved forever as a permanent Garden of Eden or ruined forever as a profitable business run by a mafia of real estate speculators. As an example of this attitude, I quote the concluding sentences of Stewart's chapter on the discovery of the islands:

> There seem to be no happy endings on Galápagos. In Chapter 7 we will find that our own times on the islands may yet prove the saddest of them all.

In chapter 7, the chapter on conservation written by Godfrey Merlen, we find the same message:

> The fact is that, in a truly fundamental way, biological equilibrium has been altered all over the world, leading to the present disaster.

If the activities of humans all over the world are considered to be a disaster, then the history of our species is indeed, as Macbeth said, a tale told by an idiot, full of sound and fury, signifying nothing. I prefer the viewpoint of Hamlet, who said, "What a piece of work is a man!"

After one week visiting the Galápagos as a tourist, I cannot claim to be an expert. But I learned enough to convince me that the black-and-white view of the islands is wrong. I learned most from our professional guides, who are obliged by law to pass examinations every few years to make sure that they know the rules of the national park as well as the names and ecological relationships of the local plants and animals. They are not required to pass examinations about the human problems of the islands, but their knowledge of the human problems is as accurate as their knowledge of the fauna and flora. To me, the human problems are the most interesting. The Galápagos National Park was established by the government of Ecuador in 1959, comprising 97 percent of the land area of the islands. Within the park, entry is tightly restricted and private ownership is prohibited.

The remaining 3 percent of the area, including the places where human settlements had already existed on four islands, remained open for private ownership and development. The 1959 division worked well for forty years. Three percent of the land was ample for a small population of settlers, most of them employed in the administration of the park and in businesses catering to the needs of tourists. In 1986 the Galápagos Marine Resources Reserve was created, adding a large area of ocean to the national park. Within the marine reserve, fishing boats and tourist boats were restricted, and large tourist boats carrying hundreds of passengers were prohibited.

Human problems became acute during the last ten years, when the number of flights between the Ecuador mainland and the airport on Baltra adjoining Santa Cruz island increased dramatically. Visitors became more numerous and also wealthier, as tourist boats and hotels became more luxurious. The province of Galápagos, with its capital at Puerto Baquerizo Moreno on San Cristóbal island, quickly became the richest province of Ecuador, and settlers poured in to share the wealth and seek their fortunes. The village of Puerto Ayora on Santa Cruz was closest to the airport and grew rapidly into a city. The price of real estate escalated.

The government found it easy to set strict limits to the numbers and movements of tourists, but it was politically impossible to set strict limits to the numbers and movements of settlers. The government could not forbid its own citizens to move from one province to another. After arriving on the islands and finding their development of property restricted by ecological rules, settlers organized protests and strikes against the local government. The park administrators now consider the most serious threat to the ecology of the islands to be the settlers and not the tourists.

Our guides described the continuing battle between libertarian settlers on the one side and environmentalist park administrators and government officials on the other. As long as both sides hope for total victory, no satisfactory solution to the problem is possible. The black-and-white view of the situation envisages only the total victory of one side or the other, either the settlers destroying the ecology or the park administrators expelling the settlers. Our guides do not believe in total victory for either side. They have their feet in both camps. They have lived their professional lives in the service of the park, educating the tourists and protecting the beauty of the islands, but they look forward to joining the private sector of the economy when they retire. They have had opportunities to invest in real estate on the islands, and the rising market will allow them an earlier and more comfortable retirement. They will do whatever they can as private citizens to spread ecological awareness among the settler population and to develop their property in an ecologically responsible fashion.

Our guides see the future of the islands as an ongoing compromise, with park administrators continuing to fight for the ecology and settlers continuing to fight for their freedom. The settlers know that their prosperity depends on the tourists who are attracted to the park, and the park administrators know that effective enforcement of their rules would be impossible without the cooperation of the provincial government elected by the settlers. The national government of Ecuador in Quito has the overall responsibility for maintaining the park and making the rules. The national government is generally in favor of a strict interpretation of the rules, but it must also be responsive to the complaints of the settlers. It is likely that the settlers will become politically stronger as their numbers and wealth increase. An enduring compromise will probably require some yielding of economic development rights from settlers to the park and some yielding of territory from the park to settlers.

In the past, the most serious damage that the settlers have done to the ecology has been the release of imported species such as goats and pigs, which multiply rapidly, run wild, and devour the native vegetation. Where the native vegetation was destroyed by goats, the population of giant tortoises also dwindled rapidly. The greatest success of the park administration has been the organized extermination of feral goats and pigs on several islands. After the exterminations, the native plants recovered quickly.

The biggest extermination campaign, getting rid of 60,000 feral goats on the biggest island, Isabela, is now coming to an end. The weapon that defeated the goats was the "Judas goats," sterilized male animals that carry radio beacons so that they can be easily located. The Judas goats attract feral goats that can then be killed by hunters. Smaller introduced species such as rats and insects can probably not be exterminated at any reasonable cost. They are on the islands to stay. The islands will never be restored completely to their original state, but the elimination of goats and pigs can cure the worst of the damage. A park kept clear of the larger feral species, including humans, is a reasonable compromise between total preservation of the ecology and total freedom for the settlers.

To achieve a stable equilibrium between the park and the settlers for centuries to come, it would be desirable to get rid of boundaries dividing individual islands. Boundaries could be redrawn so that each island is wholly in the park or wholly outside. For example, the two heavily populated islands Santa Cruz and San Cristóbal might be opened and the two lightly populated islands Isabela and Floreana might be closed to settlers. If this land swap were done now, it would require moving about 4,000 settlers out of a total of about 40,000.

The advantage of the deal for the park would be to gain complete control of the biggest island, Isabela, which is also the most diverse ecologically and geologically. The advantage for the settlers would be to possess two complete islands with a large net increase of territory. The division of land area between park and settlers would become about 85 to 15 instead of the present 97 to 3. Whether such a resettlement can be negotiated, either now or in the future, remains to be seen. The disadvantages are as obvious as the advantages: for the settlers, uprooted lives, and for the park, diminished territory. Most likely, the present division of the islands will last a long time, until some quarrel between park and settlers causes an acute crisis and compels the national government to impose a more stable arrangement.

Since I grew up in England, I tend to think of all environmental problems in terms of English analogies. England emerged out of the last ice age only 15,000 years ago, even more recently than the Galápagos emerged from the ocean, and was colonized by species migrating from the European mainland. After the newly arrived species, human settlers came to England. When they arrived, England was a pristine wilderness, and we may imagine an international park administration set up then to preserve the ecology. What should the park administration have done? What fraction of the land should have been set aside as a permanent wilderness, and what fraction should have been open to settlers? We may imagine the park administration and the settlers quarreling about these questions in England, just as they do in the Galápagos today.

In the real world, when the settlers arrived in England ten thousand years ago, there were no park administrators and no barriers to settlement. England was overrun with settlers who did what they pleased with the wilderness, first building forts on hilltops, then chopping down trees and converting forest into farmland, then building villages in valleys and cities beside rivers, then covering the country with furnaces and factories and railways and roads, polluting the air with soot and the rivers with sewage. While they were destroying the wilderness and transforming the ecology, the settlers incidentally built cathedrals and gardens, wrote plays and poems, invented machines and discovered laws of nature. Finally, in the last century, the settlers, now fifty million strong, began to clean up the environment and take care of the wildlife. Today the English countryside is entirely man-made, quite different from the original wilderness of uninterrupted forest, but it is still beautiful, rich in its variety of habitats and species. This English history raises another question. If England had been governed for the last ten thousand years by a park administration, would the final result have been better?

After examining the example of England, I do not know whether, in the long run, any international park regime dedicated to preserving the wilderness could have achieved a better result than the settlers who took possession of the land free of all restraint. Taking a long view, I am equally uncertain about the future of the Galápagos. It may well be that in the long run the settlers will do better than the park. Human settlement and wilderness are equally a part of nature, and our task as custodians of the planet is to help them to coexist peacefully. To take care of the islands in the long run, we need both the park administration and the settlers. We cannot know what problems they will face. A human presence in the islands is important not only for the future of the Galápagos but for the future of Ecuador as a whole.

After our week in the islands, we spent a second week in Ecuador, in the Amazon jungle east of the Andes. We stayed at Sacha Lodge, a tourist hotel deep in the jungle. The long trip down the river to the lodge reminded us of Conrad's *Heart of Darkness*. Our professional guides in the jungle were Amazonian natives who had grown up in that part of Ecuador. One of them had been raised by a grandfather who was a shaman, using the medicinal plants of the jungle to heal wounds and cure diseases. The grandson speaks five languages fluently: Spanish, English, and German as well as two native languages. Just as in the Galápagos, the guides are passionate environmentalists, expert in the ecology of the jungle, and also well informed about the human problems of the Amazon region. They intend to retire, as soon as they have saved enough money, to jobs in the private sector. They see the Amazon region, like the Galápagos, as a land of opportunity. For them too, ecological preservation and human presence go hand in hand.

Note added in 2014: I was disappointed to receive few responses to this review. Those that I received were mostly from people who intended to travel to the Galápagos and asked for practical advice. Nobody responded to the question that I asked about the ecological future of the islands. I still believe that I learned an important lesson from my visit, and that the lesson has important implications for people living in other parts of the planet where natural beauty is endangered by human settlement.

8

LEAPING INTO THE GRAND UNKNOWN

FRANK WILCZEK IS one of the most brilliant practitioners of particle physics. Particle physics is the science that tries to understand the smallest building blocks of earth and sky, just as biology tries to understand living creatures. Particle physics is running about two hundred years behind biology. In the eighteenth century, Carl Linnaeus started systematic biology by giving Latin names to species of plants and animals, *Homo sapiens* for humans and *Pan troglodytes* for chimpanzees. In the nineteenth century, Darwin created a unified theory for biology by explaining the origin of species. In the twentieth century, Ernest Rutherford laid the ground for particle physics by discovering that every atom has a nucleus that is vastly smaller than the atom itself, and that the nucleus is made of particles that are smaller still. In the twenty-first century, particle physicists are hoping for a new Darwin who will explain the origin of particles.

It is too soon to tell whether Wilczek will be the new Darwin. His book *The Lightness of Being** is not the new *Origin of Species*. It is more like Darwin's *Voyage of the Beagle*, a popular account of a voyage of exploration, describing the landscape and the newly discovered

***The Lightness of Being: Mass, Ether, and the Unification of Forces* (Basic Books, 2008).

creatures that still have to be explained. Wilczek is a theoretician and not an experimenter. His strength lies in leaps of the imagination rather than in heavy hardware or heavy calculations. He shared the 2004 Nobel Prize in Physics for inventing the concept that he called "asymptotic freedom."

He writes as he thinks, with a lightness of touch that can come only to one who is absolute master of his subject. He borrowed his title from Milan Kundera, the Czech writer whose novel *The Unbearable Lightness of Being* takes a gloomier view of lightness. For Wilczek, the lightness of being is not only bearable but exhilarating. He says:

> There's also a joke involved. A central theme of this book is that the ancient contrast between celestial light and earthy matter has been transcended. In modern physics, there's only one thing, and it's more like the traditional idea of light than the traditional idea of matter. Hence, *The Lightness of Being.*

Wilczek has undertaken a difficult task: to describe the central problems of particle physics to an audience ignorant of mathematics, using few equations and mostly colloquial language. His idiosyncratic jargon words, such as Core, Grid, and Jesuit Credo, are explained in an extensive glossary at the end of the book. The glossary is fun to read, full of jokes and surprises. The words Core, Grid, and Jesuit Credo are not to be found in other books about physics. They are jargon invented by Wilczek to express his personal view of the way nature works. Core is like the core curriculum that undergraduates majoring in physics are supposed to learn. It is a solidly established theory, confirmed by experiments but still obviously incomplete. It is incomplete because it describes what nature does but does not explain why. The glossary says, "The Core theory contains esthetic flaws, so we hope it is not Nature's last word."

Grid is Wilczek's word for the stuff that exists in apparently empty space. According to his view of the universe, empty space is not a featureless void. It is a highly structured, powerful medium whose activity molds the world. He says, "Where our eyes see nothing, our brains, pondering the revelations of sharply tuned experiments, discover the Grid that powers physical reality."

The Jesuit Credo refers not to a theory of the universe but to a way of approaching research: "It is more blessed to ask forgiveness than permission." This is a rule propounded by the Jesuits for saints and sinners trying to find the right way to live. If you ask for permission, the authorities will probably say no. If you ask for forgiveness, they are more likely to say yes. Wilczek was brought up in a Catholic family with a proper respect for Jesuits. The Jesuit Credo is particularly helpful for a scientist trying to find the right way to think. It is more blessed for a scientist to make a leap in the dark, and afterward be proved wrong, than to stay timidly within the limits of the known.

The main part of Wilczek's book, with the title "The Origin of Mass," describes the Core theory, the part of particle physics that is firmly based on the weak and strong forces that we observe in nature. Atoms and nuclei are held together by forces acting between all the pairs of particles that they contain. Each force acts between two particles and its strength depends on the distance between the two particles. Weak forces hold atoms together and grow weaker at large distances. Strong forces hold nuclei together and grow stronger at large distances. Large distances mean distances larger than the nucleus of an atom, and small distances mean distances smaller than a nucleus. The doctrine of asymptotic freedom, which Wilczek discovered, says that the behavior of these forces at short distances is the opposite of their behavior at large distances. At large distances, the strong force is strong and the weak force is weak, but at short distances the opposite occurs: the weak force grows stronger and the strong force grows weaker.

He called this doctrine asymptotic freedom because it implies that at high energies the strongly interacting particles become almost free. Strongly interacting particles are called hadrons, from the Greek word *hadros*, meaning fat. The higher the energy of a collision, the shorter the distance between the colliding particles. In collisions between hadrons with very high energy, the strong forces paradoxically become weak and the probabilities of collisions become small.

Another consequence of asymptotic freedom is that we can calculate the masses of hadrons, starting from a knowledge of the strength of the strong force. Masses calculated in this way agree with the observed masses of known particles. This is what Wilczek means by "The Origin of Mass." The masses of familiar objects like atoms arise from the peculiar symmetry of the strong forces. Modern theories of particle physics have the marvelous property, first pointed out by the Chinese-American physicist Frank Yang, that the strength of particle interactions is dictated by the symmetry of the theory. Since Wilczek finds the masses depending on the strength of forces, and Yang finds the strength of forces dictated by symmetry, the final result is to make mass a consequence of symmetry alone.

The last part of the book, with the title "Is Beauty Truth?," is brief and speculative. It describes a grand unified theory of particle physics going far beyond the Core, introducing a whole menagerie of hypothetical particles that are sisters to the known particles, and a symmetry principle known as supersymmetry that interchanges each known particle with its sister. The word "interchange" here does not mean a physical replacement of one particle by another. It means the mathematical interchange of the entire assemblage of known particles with the assemblage of their hypothetical sisters. The hypothesis of supersymmetry says that the equations describing the universe remain unchanged when all the known particles are interchanged with their unknown sisters. The interchange is a mathematical abstraction, not a physical action.

The grand unified theory is a bold venture into the unknown. It is a mathematical construction of spectacular beauty, unsupported by any experimental evidence. All that we can say for sure is that this theory is possibly true and certainly testable. Wilczek believes that the basic laws of nature must be beautiful, and therefore a theory that is beautiful has a good chance of being true. He believes that the grand unified theory is true because it is aesthetically pleasing. He points to several famous examples from the history of physics, when theories designed to be beautiful turned out to be true. The best-known examples are the Dirac wave equation for the electron and the Einstein theory of general relativity for gravity. If the grand unified theory turns out to be true, it will be another example of beauty lighting the way to truth.

At the end of the book, a chapter entitled "Anticipating a New Golden Age" describes Wilczek's hopes for the future of particle physics. He sees the golden age starting very soon. His hopes are based on the Large Hadron Collider (LHC), the biggest and newest particle accelerator, built by the European Center for Nuclear Research (CERN) in Geneva. The LHC is a splendid machine, accelerating two beams of particles in opposite directions around a circular vacuum pipe that has a circumference of twenty-seven kilometers. Particle detectors surround the beams where they collide, so that the products of the collisions are detected. The energy of each accelerated particle will be more than seven times the energy of a particle in any other accelerator. Wilczek confidently expects that supersymmetric sisters of known particles will be found among the debris from collisions in the LHC. By observing new particles and interactions in detail, he hopes to fill quickly the gaps in the grand unified theory.

Incidentally, Wilczek expects the LHC to solve one of the central mysteries of astronomy by identifying the dark matter that pervades the universe. We know that the universe is full of dark matter, which weighs about five times as much as the visible matter that we observe in the form of galaxies and stars. We detect the dark matter by seeing its gravitational effect on visible matter, but we do not know what the dark matter is. If the supersymmetric sisters of known particles exist, they could be the dark matter. If all goes well, the LHC will kill two birds with one stone, observing the creation of dark matter in particle collisions, and at the same time testing the theory of supersymmetry. Wilczek believes that all will go well. He sees the coming golden age as a culminating moment in the history of science:

> Through patchy clouds, off in the distance, we seem to glimpse a mathematical Paradise, where the elements that build reality shed their dross. Correcting for the distortions of our everyday vision, we create in our minds a vision of what they might really be: pure and ideal, symmetric, equal, and perfect.

Wilczek, like most scientists who are actively engaged in exploring, does not pay much attention to the history of his science. He lives in the era of particle accelerators, and assumes that particle accelerators in general, and the LHC in particular, will be the main source of experimental information about particles in the future. Since I am older and left the field of particle physics many years ago, I look at the field with a longer perspective. I find it useful to examine the past in order to explain why I disagree with Wilczek about the future. Here is a summary of the history as I remember it.

Before World War II, particle physics did not exist. We had atomic physics, the science of atoms and electrons, and nuclear physics, the science of atomic nuclei. Beyond these well-established areas of knowledge, there was a dimly lit zone of peculiar phenomena called cosmic rays. Cosmic rays were a gentle rain of high-energy particles and radiation that came down onto the earth from outer space. We called the high-energy particles mesons. Nobody knew what they were, where they came from, or why they existed. They appeared to come more or less uniformly from all directions, at all times of day or night, summer or winter. They were an enduring mystery, not yet a science.

Particle physics emerged unexpectedly in the 1940s, during the early postwar years, while the soldiers were still coming home from battlefields and prison camps. Particle physics started with makeshift equipment salvaged from the war to explore a new universe. The new field was a symbol of hope for a generation battered by war. It proved that former enemies could work together fruitfully on peaceful problems. It gave us reason to dream that friendly collaboration could spread from the world of science to the more contentious worlds of power and politics.

In 1947, Cecil Powell did a historic experiment in Bristol. He was an expert on photography and knew how to cook photographic plates so as to make them sensitive to cosmic rays. In his plates he could see tracks of cosmic rays coming to rest. When an object comes to rest in a known place at a known time, it is no longer a vague flow of unknown stuff. It is a unique and concrete object. It is accessible to the tools of science. After Powell detected a cosmic ray coming to rest, he knew where it was, and he could see what it did next. What it often did next was to produce a secondary particle moving close to the speed of light. When he started to study the secondary particles, the mystery of cosmic rays was transformed into the science of particle physics.

Powell trained an army of human image-scanners to examine with microscopes the tracks of cosmic rays coming to rest in his plates. His unique skill as an experimenter was to motivate people, not to build apparatus. His scanners worked long hours searching for rare needles in a haystack of photographic clutter. They worked together as a team. A scanner who found something new was given full credit for the discovery, but the others who worked equally hard and found nothing were given credit too. One of his scanners, Marietta Kurz, discovered a cosmic ray that came to rest twice. It stopped in a plate, then produced a secondary particle that moved a short distance before stopping again, then produced a tertiary particle that moved faster and escaped from the plate. Powell called the primary particle a pi-meson and the secondary particle a mu-meson. The pi changed into a mu, and the mu changed into something else. This experiment revealed and named the first two species in the particle zoo.

After Powell, the pioneers of particle physics continued for five years to work with cosmic rays, finding several more species of particle. One of the particles that they failed to find was the antiproton. According to theory, every particle with an electric charge should have an antiparticle with the opposite charge. The proton, which is the positively charged nucleus of the hydrogen atom and a component of every other kind of atom, should have a negatively charged twin called the antiproton. Cosmic ray experiments failed to find the antiproton because it cannot be brought to rest in matter. Every antiproton stopped in matter immediately finds a proton and annihilates itself along with its twin. Cosmic ray experts hunted for the antiproton in vain. Meanwhile, builders of particle accelerators were developing a new set of tools. Ernest Lawrence, the original inventor of the cyclotron, built a large accelerator that he called the Bevatron. In 1955 two physicists at Berkeley in California, Emilio Segrè and Owen Chamberlain, used the new accelerator to produce antiprotons in quantity and detect their annihilation. They received the Nobel Prize in 1959 for discovering the antiproton.

After 1955, a few particle physicists continued to study cosmic rays and other kinds of natural radiation with passive detectors, but the new experimental tool, the high-energy accelerator, rapidly took over the field. Particle accelerators had many advantages over passive detectors. Accelerators provided particles in far greater numbers, with precisely known energies, under the control of the experimenter. Accelerator experiments were more quantitative and more precise. But accelerators also had some serious disadvantages. They were more expensive than passive detectors, they required teams of engineers to keep them running, and they produced particles with a limited range of energies.

Nature provided among the cosmic rays a small number of particles with energies millions of times larger than the largest accelerator could reach. If the distribution of effort between accelerators and passive detectors had been rationally planned, particle physicists would have maintained a balance between the two types of instrument, perhaps three quarters of the money for accelerators and one quarter for passive detectors. Instead, accelerators became the prevailing fashion. The era of accelerator physics had begun, and big accelerators became political status symbols for countries competing for scientific leadership. For forty years after 1955, the United States built a succession of big accelerators and only two passive detectors. The Soviet Union and CERN followed suit, putting almost all their efforts into accelerators. Meanwhile, serious research using passive detectors continued in Canada and Japan, countries with high scientific standards and limited resources.

In the United States, Raymond Davis Jr. was a lonely pioneer who found a new way of doing experiments with natural radiation. He demonstrated that he could detect the appearance of a single atom of argon in a tank containing six hundred tons of a common industrial cleaning fluid. This cleaning fluid is cheap and available in big quantities. It consists of 13 percent carbon and 87 percent chlorine. Argon is a gas with properties totally different from chlorine. Davis put his tank full of cleaning fluid a mile underground in a mined-out cavity belonging to the Homestake gold mine in South Dakota, so as to get away from the confusing effects of cosmic rays. He was interested in observing natural radiation from the center of the sun. According to the standard model of nuclear energy generation in the sun, the sun produces particles called neutrinos, which arrive at the earth and very rarely cause chlorine atoms to change into argon atoms. The predicted rate of appearance of argon atoms in Davis's tank was three per month. Davis claimed that he could reliably count the argon atoms. He counted them for many years and found only one per month instead of three. The deficiency of argon atoms was known as the "solar neutrino problem."

The solar neutrino problem could be explained in three ways. Either Davis's experiment was wrong, or the standard model of the sun was wrong, or the standard theory of the neutrino was wrong. For many years, most of the experts believed that the experiment was wrong, that Davis missed two thirds of the argon atoms because they slipped through his counters. Davis did some careful tests that convinced the experts that his counters were not to blame, and then they mostly believed that the model of the sun was wrong. The model of the sun was checked by accurate measurements of seismic waves traveling through the sun, and turned out to be correct. So the experts finally had to admit that their theory of the neutrino was wrong.

We now know that there are three kinds of neutrinos. Only one kind is produced in the sun, and only that kind was detected in Davis's tank, but many switch smoothly from one kind to another while they are traveling from the sun to the earth. Two thirds of them are the wrong kind to be detected when they arrive at the tank, neatly explaining Davis's result. This discovery was the first evidence for processes not included in the scheme that Wilczek calls the Core. Davis was awarded a belated Nobel Prize for it in 2002. During the years while Davis was working alone with his tank, larger teams of physicists and engineers were making discoveries at a rapid pace with accelerators. The accelerator era was in full swing. Particle physics as we know it today is largely the fruit of accelerators.

So much for the history. Now I turn from the past to the future. Wilczek's expectation, that the advent of the LHC will bring a golden age of particle physics, is widely shared among physicists and widely propagated in the press and television. The public is led to believe that the LHC is the only road to glory. This belief is dangerous because it promises too much. If it should happen that the LHC fails, the public may decide that particle physics is no longer worth supporting. The public needs to hear some bad news and some good news. The bad news is that the LHC may fail. The good news is that if the LHC fails, there are other ways to explore the world of particles and arrive at a golden age. The failure of the LHC would be a serious setback, but it would not be the end of particle physics.

There are two reasons to be skeptical about the importance of the LHC: one technical and one historical. The technical weakness of the LHC arises from the nature of the collisions that it studies. These are collisions of protons with protons, and they have the unfortunate habit of being messy. Two protons colliding at the energy of the LHC behave rather like two sandbags, splitting open and strewing sand in all directions. A typical proton–proton collision in the LHC will produce a large spray of secondary particles, and the collisions are occurring at a rate of millions per second. The machine must automatically discard the vast majority of the collisions, so that the small minority that might be scientifically important can be precisely recorded and analyzed. The criteria for discarding events must be written into the software program that controls the handling of information. The software program tells the detectors which collisions to ignore. There is a serious danger that the LHC can discover only things that the programmers of the software expected. The most important discoveries may be things that nobody expected. The most important discoveries may be missed.

Another way to go ahead with particle physics is to follow the lead of Davis and build large passive detectors observing natural radiation. In the last twenty years, the two most ambitious passive detectors were built in Canada and Japan. Both of these detectors made important discoveries, confirming and completing the work of Davis. In a well-designed passive detector deep underground, events of any kind are rare, every event is recorded in detail, and if anything unexpected happens you will see it.

There are also historical reasons not to expect too much from the LHC. I have made a survey of the history of important discoveries in particle physics over the last sixty years. To avoid making personal judgments about importance, I define an important discovery to be one that resulted in a Nobel Prize for the discoverers. This is an objective criterion, and it usually agrees with my subjective judgment. In my opinion, the Nobel Committee has made remarkably few mistakes in its awards. There have been sixteen important experimental discoveries between 1945 and 2008.

Each experimental discovery lies on one of three frontiers between known and unknown territory. It is on the energy frontier if it reaches a new range of energy of particles. It is on the rarity frontier if it reaches a new range of rarity of events. It is on the accuracy frontier if it reaches a new range of accuracy of measurements. I assigned each of the sixteen important discoveries to one of the three frontiers. In most cases, the assignments are unambiguous. For example, two of the three discoveries that I mentioned earlier, Powell's discovery of double-stopping mesons and Davis's discovery of missing solar neutrinos, lie on the rarity frontier, while only one, Segrè and Chamberlain's discovery of the antiproton, lies on the energy frontier.

The results of my survey are then as follows: four discoveries on the energy frontier, four on the rarity frontier, eight on the accuracy frontier. Only a quarter of the discoveries were made on the energy frontier, while half of them were made on the accuracy frontier. For making important discoveries, high accuracy was more useful than high energy. The historical record contradicts the prevailing view that the LHC is the indispensable tool for new discoveries because it has the highest energy.

The majority of young particle physicists today believe in big accelerators as the essential tools of their trade. Like Napoleon, they believe that God is on the side of the big battalions. They consider passive detectors of natural radiation to be quaint relics of ancient times. When I say that passive detectors may still beat accelerators at the game of discovery, they think this is the wishful thinking of an old man in love with the past. I freely admit that I am guilty of wishful thinking. I have a sentimental attachment to passive detectors, and a dislike of machines that cost billions of dollars to build and inevitably become embroiled in politics. But I see evidence, in the recent triumphs of passive detectors and the diminishing fertility of accelerators, that nature may share my prejudices. I leave it to nature to decide whether passive detectors or the LHC will prevail in the race to discover her secrets.

Fortunately, passive detectors are much cheaper than the LHC. The best of the existing passive detectors were built by Canada and

Japan, countries that could not afford to build giant accelerators. The race for important discoveries does not always go to the highest energy and the most expensive machine. More often than not, the race goes to the smartest brain. After all, that is why Wilczek won a Nobel Prize.

Note added in 2014: To be fair to Wilczek, I include his response to the review.

> I know you'd be disappointed if I agreed with everything you said, so I'll append my response to someone who asked about it: Although I enjoyed Dyson's review, a few points of his seemed off to me. For instance, I don't think it's reasonable to compare particle physics today to biology before Darwin. In fundamental physics we have very sophisticated, specific, successful mathematical theories, of a kind that biologists can barely dream of even today. As to "active versus passive," I don't think it's an either/or proposition. Different questions call for different methods of investigation. Without going into technicalities, here's a short list of central open questions that are more suitable for non-accelerator versus accelerator physics. Non-accelerator: proton decay, intrinsic electric dipole moments, dark matter annihilation signatures, axion or other ultra-light particle searches. Accelerator: Higgs sector, supersymmetry, production of dark matter candidate particles, surprises in high energy interactions.
>
> With all best wishes,
> Frank W.

9

WHEN SCIENCE AND POETRY WERE FRIENDS

THE AGE OF WONDER means the period of sixty years between 1770 and 1830, commonly called the Romantic Age. It is most clearly defined as an age of poetry. As every English schoolchild of my generation learned, the Romantic Age had three major poets, Blake and Wordsworth and Coleridge, at the beginning, and three more major poets, Shelley and Keats and Byron, at the end. In literary style it is sharply different from the Classical Age before it (Dryden and Pope) and the Victorian Age after it (Tennyson and Browning). Looking at nature, Blake saw a vision of wildness:

Tyger, tyger, burning bright,
In the forests of the night;
What immortal hand or eye,
Could frame thy fearful symmetry?

Byron saw a vision of darkness:

The bright sun was extinguish'd, and the stars
Did wander darkling in the eternal space,

Rayless, and pathless, and the icy earth
Swung blind and blackening in the moonless air....

During the same period there were great Romantic poets in other countries, Goethe and Schiller in Germany and Pushkin in Russia, but in his book *The Age of Wonder*, Richard Holmes writes only about the local scene in England.*

Holmes is well known as a biographer. He has published biographies of Coleridge and Shelley and other literary heroes. But this book is primarily concerned with scientists rather than with poets. The central figures in the story are the botanist Joseph Banks; the chemists Humphry Davy and Michael Faraday; the astronomers William Herschel and his sister, Caroline, and son, John; the medical doctors Erasmus Darwin and William Lawrence; and the explorers James Cook and Mungo Park. The scientists of that age were as romantic as the poets. The scientific discoveries were as unexpected and intoxicating as the poems. Many of the poets were intensely interested in science, and many of the scientists in poetry.

The scientists and the poets belonged to a single culture and were in many cases personal friends. Erasmus Darwin, the grandfather of Charles Darwin and progenitor of many of Charles's ideas, published his speculations about evolution in a book-length poem, *The Botanic Garden*, in 1791. Davy wrote poetry all his life and published much of it. He was a close friend of Coleridge, Shelley a close friend of Lawrence. The boundless prodigality of nature inspired scientists and poets with the same feelings of wonder. *The Age of Wonder* is popular history at its best, racy, readable, and well documented. The

**The Age of Wonder: How the Romantic Generation Discovered the Beauty and Terror of Science* (Pantheon, 2008).

subtitle, "How the Romantic Generation Discovered the Beauty and Terror of Science," accurately describes what happened.

Holmes presents the drama in ten scenes, each dominated by one or two of the leading characters. The first scene belongs to Joseph Banks, who sailed with Captain James Cook on the ship *Endeavour.* This was Cook's first voyage around the world. One of the purposes of the expedition was to observe the transit of Venus across the disk of the sun on June 3, 1769, from the island of Tahiti in the South Pacific. The tracking of the transit from the Southern Hemisphere, in combination with similar observations made from Europe, would give astronomers more accurate knowledge of the distance of the earth from the sun. Banks was officially the chief botanist of the expedition, but he quickly became more interested in the human inhabitants of the island than in the plants. The ship stayed for three months at Tahiti, and he spent most of the time, including the nights, ashore. During the nights he was not observing plants.

A wealthy young man accustomed to aristocratic privileges in England, Banks quickly made friends with the Tahitian queen Oborea, who assigned one of her personal servants, Otheothea, to take care of him. With the help of Otheothea and other good friends, he acquired some fluency in the Tahitian language and customs. His journal contains a Tahitian vocabulary and detailed descriptions of the people he came to know. When the time came to set up the astronomical instruments and observe the transit of Venus, he took the trouble to explain to his Tahitian friends what was happening. "To them we shewd the planet upon the sun and made them understand that we came on purpose to see it."

During the long months at sea after leaving Tahiti, Banks rewrote his journal entries into a formal narrative, "On the Manners and Customs of the South Sea Islands," one of the founding documents of the science of anthropology. In a less formal essay written after his return to England, he wrote:

> In the Island of Otaheite where Love is the Chief Occupation, the favourite, nay almost the Sole Luxury of the Inhabitants, both the bodies and souls of the women are modeld in the utmost perfection for that soft science.

The Tahiti that he describes was truly an earthly paradise, not yet ravaged by European greed and European diseases, twenty years before the visit of William Bligh and the *Bounty* mutineers, sixty-six years before the visit of Charles Darwin and the *Beagle*.

After exploring the South Seas, Cook sailed down the eastern coast of Australia and landed at Botany Bay. Banks failed to establish social contacts with the Australian aborigines and returned to his role as botanist, bringing back to England a treasure trove of exotic plants, many of them today carrying his name. After he returned to England, he found that he and Captain Cook had become public heroes. He was invited to meet King George III, who was then young and of sound mind and shared his passion for botany. He remained a lifelong friend of the king, who actively supported his creation of the national botanic garden at Kew.

Banks became the president of the Royal Society in 1778 and held that office for forty-two years, officially presiding over British science for more than half of the Age of Wonder. He presided with a light hand and did not attempt to turn the Royal Society into a professional organization like the academies of science in Paris and Berlin. He believed that science was best done by amateurs like himself. If some financial support was needed for people without private means, it could best be provided by aristocratic patrons.

One of those for whom Banks found support was William Herschel, the greatest astronomer of the age. Herschel was a native Hanoverian, and was conscripted at the age of seventeen to fight for Hanover in the Seven Years' War against the French. After surviving a battle that the Hanoverians lost, he escaped to England to begin a new life as a professional musician. Starting as a penniless refugee, he rose rapidly in the English musical world. By his late twenties he was the director of the orchestra in the Pump Room at Bath, the health resort where people of wealth congregated to take the waters and listen to concerts. He stayed at Bath for sixteen years, running the musical life of the city by day and scanning the sky at night. As an astronomer he was a complete amateur, unpaid and self-taught.

At the beginning, when Herschel began observing the heavenly bodies, he believed that they were inhabited by intelligent aliens. The round objects that he saw on the moon were cities that the aliens had built. He continued throughout his life to publish wild speculations, many of which turned out later to be correct. He had two great advantages as an observer. First, he built his own instruments, and with his musician's hands made exquisitely figured mirrors that gave sharper images than any other telescopes then existing. Second, he brought his younger sister, Caroline, over from Hanover to be his assistant, and she became an expert observer with many independent discoveries to her credit. His life as an amateur ended in 1781 when with Caroline's help he discovered the planet Uranus.

As soon as Banks heard of the discovery, he invited Herschel to dinner, introduced him to the king, and arranged for him to be appointed the king's personal astronomer with a salary of £200 a year, later supplemented by a separate salary of £50 a year for Caroline. Herschel's musical career was over, and he spent the rest of his life as a professional astronomer. He obtained royal funding to build bigger telescopes, and embarked on a systematic survey of every star and nebulous object in the sky, pushing his search outward to include objects fainter and more distant than anyone else had seen.

Herschel understood that when he looked at remote objects he was looking not only into deep space but into deep time. He correctly identified many of the nebulous objects as external galaxies like our own Milky Way, and calculated that he was seeing them as they existed at least two million years in the past. He showed that the universe was not only immensely large but immensely old. He published papers that moved away from the old Aristotelian view of the heavens as a static domain of perpetual peace and harmony, and toward the modern view of the universe as a dynamically evolving system. He wrote of "a gradual dissolution of the Milky Way" that would provide "a kind of chronometer that may be used to measure the time of its past and future existence." This idea of a galactic chronometer was the beginning of the new science of cosmology.

As Holmes's account suggests, all the leading scientists of the Romantic Age, like Banks and Herschel, started their lives as brilliant, unconventional, credulous, and adventurous amateurs. They blundered into science or literature in pursuit of ideas that were often absurd. They became sober professionals only after they had achieved success. Another example was Humphry Davy, who originally intended to be a physician and worked, as part of his medical training, as an assistant at the Pneumatic Institution in Bristol. The Pneumatic Institution was a clinic where patients were treated for ailments of all kinds by inhaling gases. Among the gases available for inhaling was nitrous oxide. Davy experimented enthusiastically with nitrous oxide, using himself and his friends, including Coleridge, as subjects. After one of these sessions, he wrote:

> I have felt a more high degree of pleasure from breathing nitrous oxide than I ever felt from any cause whatever—a thrill-

ing all over me most exquisitely pleasurable, I said to myself I was born to benefit the world by my great talents.

Davy was so popular in Bristol that he was invited at the age of twenty-three to become an assistant lecturer in chemistry at the Royal Institution in London. The Royal Institution was a newly founded venture that provided "regular courses of philosophical lectures and experiments" for fashionable London audiences. For the preparation of experimental demonstrations to astound and educate the public, the lecturer was provided with a laboratory where he could also do original research.

Davy promptly switched his research activities from physiology to chemistry. He became the first electrochemist, using a huge electric battery to decompose chemical compounds, and discovered the elements sodium and potassium. Later he invented the Davy safety lamp, which made it possible for coal miners to work underground without killing themselves in methane explosions. The lamp made him even more famous. Coleridge invited him to move north and establish a chemical laboratory in the Lake District where Coleridge and Wordsworth lived. Coleridge wrote to him, "I shall attack Chemistry like a Shark." Davy wisely stayed in London, where he succeeded Banks as the president of the Royal Society and chief panjandrum of British science. Byron gave him a couple of lines in his poem *Don Juan*:

This is the patent-age of new inventions
For killing bodies, and for saving souls,
All propagated with the best intentions;
Sir Humphry Davy's lantern, by which coals
Are safely mined for in the mode he mentions,
Tombuctoo travels, voyages to the Poles,

Are ways to benefit mankind, as true,
Perhaps, as shooting them at Waterloo.

The question that Byron raised, whether scientific advances and inventions truly benefit mankind, was answered dramatically in the negative by Victor Frankenstein, one of the most durable creations of the Age of Wonder. Mary Shelley, wife of the poet, was nineteen years old in 1817 when she wrote her novel *Frankenstein, or the Modern Prometheus.* In the same year, her husband was frequently visiting the physician William Lawrence, both as a patient and a close friend. Lawrence wrote a popular book, *Lectures on the Natural History of Man*, a scientific account of human anatomy and physiology, based on recent discoveries by surgeons in dissecting rooms. Lawrence fiercely attacked the doctrine of vitalism that was then fashionable. According to the vitalists, there exists a life force that animates living creatures and makes them fundamentally different from dead matter. Lawrence was a materialist and believed in no such force. Holmes discusses the question whether Mary's idea for her novel arose from the intellectual battle between vitalists and materialists or from the actual attempts of the notorious charlatans Giovanni Aldini in England and Johann Ritter in Germany to revive dead animals with electric currents. Aldini had on one occasion publicly attempted to revive the corpse of a human murderer.

The novel portrays Frankenstein creating his monster silently by candlelight, using the delicate dissecting tools of a surgeon, and portrays the monster as an articulate philosopher lamenting his loneliness in poignantly poetic language. Six years later, the novel was turned into a play, *Presumption: or the Fate of Frankenstein*, which was a big success in London, Bristol, Paris, and New York. The play turned Mary Shelley's intellectual drama upside-down. It became a combination of horror story with black comedy, and that is the way it has remained ever since, on the stage and in the movies. In the play, the monster is created by zapping dead flesh with sparks from a huge electrical machine, and the creature emerges as a dumb and misshapen caricature of a human, the epitome of brutal malevolence. And then comes a surprise. Mary went to see the play and loved it. She wrote in a letter to a friend:

> Lo and behold! I found myself famous! . . . Mr. Cooke played the "*blank's*" part extremely well . . . all he does was well imagined and executed . . . it appears to excite a breathless excitement in the audience . . . in the early performances all the ladies fainted and hubbub ensued!

She called the monster "blank" because its name was left blank in the theater program. She was seduced by the magic of show business in 1823, just as young writers are seduced by the magic of show business today.

In 1831 Mary Shelley wrote a preface for a new edition of the novel. The preface describes her memories of the genesis of her masterpiece fourteen years earlier:

> I saw the pale student of unhallowed arts kneeling beside the Thing he had put together. I saw the hideous phantasm of a man stretched out, and then, on the working of some powerful Engine, show signs of life and stir with an uneasy, half-vital motion. Frightful must it be, for supremely frightful would be the effect of any human endeavour to mock the stupendous mechanism of the Creator of the world.

Her memories were closer to the play than to the novel. In Mary's original conception, the monster was a character capable of happiness and unselfish love, who turned to evil only when Frankenstein refused to create a female partner for it to love and cherish. But on the stage and ever afterward, it became pure evil, an unmitigated disaster. Science became not merely ethically ambiguous but an agent of doom.

The Age of Wonder, according to Holmes, ended with the first meeting of the British Association for the Advancement of Science (BAAS) in York in 1831. By that time the three giants, Banks, Herschel, and Davy, had grown old and feeble and finally died. The three young leaders who took their places were the mathematician Charles Babbage, the astronomer John Herschel, and the physicist David Brewster. Babbage led the attack on the old regime in 1830 with a book, *Reflections on the Decline of Science in England*. He attacked the dignitaries of the Royal Society in London as a group of idle and incompetent snobs, out of touch with the modern world of science and industry. The professional scientists of France and Germany were leaving the English amateurs far behind. England needed a new organization of scientists, based in the growing industrial cities of the north rather than in London, run by active professionals rather than by gentleman amateurs. The BAAS was set up according to Babbage's specifications, with annual meetings held in various provincial cities but never in London. Membership grew rapidly. At the third meeting in Cambridge in 1833, the word "scientist" was used for the first time instead of "natural philosopher," to emphasize the break with the past. Victoria was not yet queen, but the Victorian Age had begun.

Holmes's history of the Age of Wonder raises an intriguing question about the present age. Is it possible that we are now entering a new Romantic Age, extending over the first half of the twenty-first century, with the technological billionaires of today playing roles similar to the enlightened aristocrats of the eighteenth century? It is too soon now to answer this question, but it is not too soon to begin examining the evidence. The evidence for a new Age of Wonder would be a shift backward in the culture of science, from organizations to individuals, from professionals to amateurs, from programs of research to works of art.

If the new Romantic Age is real, it will be centered on biology and computers, as the old one was centered on chemistry and poetry. Candidates for leadership of the modern Romantic Age are the biology wizards Kary Mullis, Dean Kamen, and Craig Venter, and the computer wizards Larry Page, Sergey Brin, and Charles Simonyi. Venter is the entrepreneur who taught the world how to read genomes fast; Mullis is the surfer who taught the world how to multiply genomes fast; Kamen is the medical engineer who taught the world how to make artificial hands that really work.

Each achievement of our modern pioneers resonates with echoes from the past. Venter sailed around the world on his yacht collecting genomes of microbes from the ocean and sequencing them wholesale, like Banks who sailed around the world collecting plants. Mullis invented the polymerase chain reaction, which allows biologists to multiply a single molecule of DNA into a bucketful of identical molecules within a few hours, and after that spent most of his time surfing the beaches of California, like Davy who invented the miners' lamp and after that spent much of his time fly-fishing along the rivers of Scotland.

Kamen builds linkages between living human brains and mechanical fingers and thumbs, like Victor Frankenstein, who sewed dead brains and hands together and brought them to life. Page and Brin built the giant Google search engine that reaches out to the furthest limits of human knowledge, like William Herschel, who built his giant forty-foot telescope to reach out to the limits of the universe. Simonyi was the chief architect of software systems for Microsoft and later flew twice as a cosmonaut on the International Space Station, like the intrepid aeronauts Pierre Blanchard and John Jeffries, who made the first aerial voyage from England to France by balloon in 1785.

There are obvious differences between the modern age and the Age of Wonder. Now we have a standing army of many thousands of professional scientists. Then we had only a handful. Now science has become an organized professional activity with big budgets and big payrolls. Then science was a mixture of private hobbies and public entertainments. In spite of the differences, there are many similarities. Holmes remarks that in 1812 "Portable Chemical Chests" began to go on sale in Piccadilly, priced between six and twenty guineas. These contained equipment and materials for serious amateur chemists.

Their existence proves that some of the fashionable ladies and gentlemen who swarmed to Davy's public lectures at the Royal Institution either did real chemical experiments in their homes or encouraged their children to do such experiments. Last year I received as a Christmas present a "Portable Genome Chest," a CD containing a substantial amount of information about my genome. My children and grandchildren, and our spouses, got their CDs too. By comparing our genomes, we can measure quantitatively how much each grandchild inherited from each grandparent.

The CDs tell us the places where our personal DNA differs from the standard human genome by a single letter of the genetic code. Other more complicated differences, such as deletions or repetitions of a string of letters, are not included. The CDs are prepared and sold by a company called 23andMe, twenty-three being the number of chromosomes in a human germ cell. The founder of the company is Anne Wojcicki, the wife of Brin.

The language of the genome is still an undeciphered script, like the Linear B script after it was discovered on ancient clay tablets in Crete. Professional archaeologists and linguists failed for fifty years to decipher Linear B. The amateur Michael Ventris succeeded where the experts had failed, and proved that Linear B was a pre-Homeric form of Greek. I am certainly no Ventris. I cannot decipher my own genome, or extract from it any useful information about my anatomy and physiology. But I consider it a cause for celebration that personal genetic information is now widely distributed at a price that ordinary citizens can afford. Before long, complete human genomes will also be widely available. Then we will see whether the professional experts will win the race to understand the subtle architecture of the genome, or whether some new Ventris will beat them at their own game.

An important step toward an understanding of the genome is the recent work of David Haussler and his colleagues at the University of California at Santa Cruz, published in the online edition of *Nature*.* Haussler is a professional computer expert who switched his interest to biology. He never dissected a cadaver of a mouse or a human. His experimental tool is an ordinary computer, which he and his students use to make precise comparisons of genomes of different species. They discovered a small patch of DNA in the genome of vertebrates that has been strictly conserved in the genomes of chickens, mice, rats, and chimpanzees, but strongly modified in humans. The patch is called HAR1, short for human accelerated region 1. It evolved hardly at all in three hundred million years from the common ancestor of chickens and mice to the common ancestor of chimpanzees and humans, and then evolved rapidly in six million years from the common ancestor of chimpanzees and humans to modern humans.

During the last six million years, eighteen changes became fixed in this patch of the human germ line. Some major reorganization must

*"An RNA Gene Expressed During Cortical Development Evolved Rapidly in Humans," *Nature* (online), August 16, 2006.

have occurred in the developmental program that this patch helps to regulate. Another crucial fact is known about HAR1. It is active in the developing cortex of the embryo brain during the second trimester of the mother's pregnancy, the time when the detailed structure of the brain is organized. Haussler's team found another similar patch of DNA in the vertebrate genome, which they call HAR2. It is active in the developing wrist of the human embryo hand. The brain and the hand are the two organs that most sharply differentiate humans from our vertebrate cousins.

The discovery of HAR1 and HAR2 is probably an event of seminal importance, comparable with the discovery of the nucleus of the atom by Ernest Rutherford in 1909 or the discovery of the double helix in the nucleus of the cell by Francis Crick and James Watson in 1953. It opens the door to a new science, the study of human nature at the molecular level. This new science will profoundly change the possible applications of biological knowledge for good or evil. It may give us the key to control the evolution of our own species.

One feature of the old Age of Wonder is conspicuously absent in the new age. Poetry, the dominant art form in many human cultures from Homer to Byron, no longer dominates. No modern poet has the stature of Coleridge or Shelley. Poetry has in part been replaced in the popular culture by graphic art. In 2008 I took part in a "Festival of Mathematics" organized in Rome by Piergiorgio Odifreddi, a mathematical entrepreneur in tune with the modern age. Odifreddi borrowed the largest auditorium in Rome, left over from the 1960 Olympic Games, and filled every seat for three days with young people celebrating mathematics. How did he do it? By mixing mathematics with art. The presiding geniuses were the late artist Maurits Escher and the mathematician Benoit Mandelbrot, with their followers displaying new works of art created by humans and computers.

John Nash was there, enjoying the adulation of the students since the film *A Beautiful Mind* made him an international star. There was also a performing juggler who happens to be a professor of mathematics. He stood on the stage, simultaneously juggling five balls and proving elegant theorems about the combinatorics of juggling. His theorems explain why serious jugglers always juggle with an odd number of balls, usually five or seven rather than four or six.

If the dominant science in the new Age of Wonder is biology, then the dominant art form should be the design of genomes to create new varieties of animals and plants. This art form, using the new biotechnology creatively to enhance the ancient skills of plant and animal breeders, is still struggling to be born. It must struggle against cultural barriers as well as technical difficulties, against the myth of Frankenstein as well as the reality of genetic defects and deformities.

If this dream comes true, and the new art form emerges triumphant, then a new generation of artists, writing genomes as fluently as Blake and Byron wrote verses, might create an abundance of new flowers and fruit and trees and birds to enrich the ecology of our planet. Most of these artists would be amateurs, but they would be in close touch with science, like the poets of the earlier Age of Wonder. The new Age of Wonder might bring together wealthy entrepreneurs like Venter and Kamen, academic professionals like Haussler, and a worldwide community of gardeners and farmers and breeders, working together to make the planet beautiful as well as fertile, hospitable to hummingbirds as well as to humans.

Note added in 2014: Since this review was written, several private ventures have launched rockets into orbit, competing successfully with government programs. Elon Musk, the billionaire founder of the SpaceX company, is leading this new wave of amateur space pioneers. Their dream is to create a new space industry, opening the heavens to private explorers and settlers. They maintain friendly relationships with the professional space scientists and engineers who share their vision.

10

WHAT PRICE GLORY?

STEVEN WEINBERG IS famous as a scientist, but he thinks deeply and writes elegantly about many other things besides science. *Lake Views: This World and the Universe*, a collection of his writings, is concerned with history, politics, and science in roughly equal measure.* The picture on the jacket shows dark waves on deep water with a distant suburban shoreline. The water is Lake Austin in Texas, and the picture is a view taken from the window of the study where Weinberg thinks and writes. He is a native of New York who has taken root and flourished in Texas. His chief contribution to our civilization is his leadership in the understanding of nature. After twenty years of experiments in particle physics had displayed a tangled landscape of particles interacting with one another in incomprehensible ways, Weinberg's mathematical wizardry dispelled the confusion and revealed an underlying unity.

He is not only preeminent as a mathematical physicist. He has also made important contributions to the discussion of history and politics. He is one of the founders of the Union of Concerned Scientists, a group of citizens who have worked steadily for forty years to bring

*Belknap Press/Harvard University Press, 2009.

scientific wisdom into public debates about political and military problems. He has been called to Washington to testify at congressional committee hearings on strategic questions. He has become almost as expert in military history as he is in mathematical physics.

A reader who has time for only one piece should read chapter 12, "What Price Glory?" It goes deeply into the history of military technology, from the twenty-first century all the way back to the eleventh. Weinberg finds in many diverse times and places a common theme. Military leaders and military institutions have a constant tendency to glorify technology that is colorful and spectacular, even when it leads them repeatedly to defeat and disaster.

The most durable of the glorified technologies was the medieval horse carrying a knight in armor. The knight was armed with a heavy lance pointing forward. The tactic by which the knight was supposed to win battles was a cavalry charge, the horses and lances overwhelming foot soldiers with irresistible force. Weinberg examines the evidence and finds that successful cavalry charges were rare. More often, foot soldiers defeated the charge by moving out of the way of the horses or by occupying strong defensive positions. After the charge was over and the knights were dispersed, foot soldiers could defeat them individually by force of numbers.

Weinberg describes with scholarly relish several historic battles in which foot soldiers defeated cavalry. In spite of these repeated calamities, the knight on his horse remained the emblem of military virtue throughout the long centuries of the Middle Ages. Kings and emperors spent their fortunes and gave land to their feudal dependents to pay for knights and horses. In times of peace, the knights and horses exercised their military skills by competing with one another in splendid tournaments. The display of fine armor and equestrian skill became an end in itself, pursued by knights and armorers without much regard for military effectiveness. Making a grand spectacle in tournaments was more important than winning battles against peasants armed with bows and arrows. According to the customs of the Middle Ages, a knight who survived a defeat by peasants could usually return home without dishonor, after paying a ransom appropriate to his rank in the feudal hierarchy. The ransom might ruin his feudal estate but would not ruin his military career.

Weinberg finds a historic continuity between the medieval hero worship of the knight on horseback and the modern hero worship of men riding invincible machines. He tells the sad story of the *Dreadnought*, a British battleship that was launched in 1905, the brainchild of Sir John Fisher. Fisher was an anomaly in the Royal Navy, a technical expert who had risen to the rank of First Sea Lord. Unlike other Sea Lords, he understood the technologies of gunnery and torpedoes. The *Dreadnought* was the fulfillment of his dream to build a ship that was technologically supreme, faster and more heavily armed than any ship afloat, outrunning and outgunning any possible opponent.

In 1905 Britannia still ruled the waves. My father was then at the Royal Naval College at Osborne, teaching music to naval cadets. The sons of the British aristocracy were trained at Osborne to become officers of the Royal Navy, just as they had been trained in past centuries to become warriors on horseback. My father had two future kings, Edward VIII and George VI, singing in his boy choir. He enjoyed the welcoming message emblazoned on a banner above the main entrance to the college: "There is nothing the Royal Navy cannot do." He shared the prevalent admiration for the Royal Navy and for Sir John Fisher. The Royal Navy was then bigger and stronger than any two other navies combined. Without the *Dreadnought*, Britain would probably have stayed ahead of other navies for a long time.

But the *Dreadnought* attracted intense public attention all over the world. It made the rest of the Royal Navy seem suddenly worthless. After 1905, the only ships that counted politically were dreadnoughts.

The Kaiser decided to build dreadnoughts, and Britain had to build more dreadnoughts to keep up with Germany. A serious naval arms race had begun, which continued with increasing intensity until World War I broke out in 1914. The dreadnoughts had destroyed Britain's naval superiority. They made the British and German navies appear to be equal. They may have helped significantly to upset the political balance of Europe and to precipitate the tragedy of 1914.

After World War I began, the British and German dreadnoughts did not play an important part. They engaged each other only once, at the inconclusive Battle of Jutland in 1916. The decisive struggle between the two navies began in 1917, when German submarines almost cut off Britain's vital war supplies by destroying vast numbers of merchant ships, and British destroyer escorts barely succeeded in sinking enough submarines so that convoys of merchant ships could survive. If the Kaiser had built more submarines instead of dreadnoughts, he might have won the war, and if the British had built more destroyers they might have won it sooner. But the mystique of the invincible battleship lasted through World War I and afterward, until it was finally demolished by the victories of aircraft carriers in World War II.

Already in World War I, the myth of the battleship was being displaced by the new mythology of air power. The first airman hero was the Red Baron, the German Manfred von Richthofen, who flew his red triplane flamboyantly over the western front. After a year and a half of spectacular combat, he was killed, but he remained a durable hero. Meanwhile Hugh Trenchard, a less flamboyant but more important airman hero, emerged on the British side. Trenchard was the commander of the Royal Flying Corps, at that time subordinate to the army and engaged in tactical operations in France. Trenchard flew low over the western front and saw with his own eyes the miseries of the soldiers in their muddy trenches. He had dreams of a different kind of war, in which the agonies of the western front would be avoided. In his dreams he would fly his airplanes not to France but to Germany. He would bring the war to the German homeland and win it there. His airmen would destroy the German war industries without help from the army. They would attack the German war leaders directly in Berlin and save the lives of the millions of young men in the trenches.

After World War I ended, Trenchard turned his dreams into reality. The Royal Flying Corps became the Royal Air Force, with independent authority to fight its own wars. Trenchard remained in command. When Hitler came to power in Germany, Britain had to prepare for World War II with a serious program of rearmament. Trenchard had retired as chief of the Royal Air Force, but his views prevailed. The decision was made in 1936 that Britain's primary instrument for fighting the next war would be the Royal Air Force Bomber Command. Bomber Command would be a huge force of heavy bombers, designed for the strategic bombing of Germany and not for tactical support of the army. Never again would Britain fight a war in trenches. Hitler would be defeated in the air over Berlin. At the same time, while these decisions were being implemented in Britain, similar decisions were made in the US, with the airman Billy Mitchell playing the role of Trenchard. The US also believed in victory through air power and built a large force of strategic bombers.

Hitler did not believe in strategic bombers. Neither did the leaders of Japan and the Soviet Union. As a result, strategic bombers had a minor part in the outcome of World War II. I was at the headquarters of Bomber Command in the winter of 1943–1944 when we launched the series of sixteen massive attacks on Berlin that were intended to "knock Hitler out of the war" without the unpleasantness of invading France. Bomber Command had finally grown to the size that the planners of 1936 had specified. The crews and the machines were ready for action. This was our opportunity to win the war in the air over Berlin, as Trenchard had imagined in 1917. We failed miserably. Persistent winter clouds over the city made accurate bombing impossible. Three thousand of our young men died in the attacks. Our losses of bombers grew heavier as the German fighters improved their skills and tactics.

The population of Berlin kept the city functioning, and the German war industries continued to increase their production. By January 1944 it was clear to us at Bomber Command headquarters that the Battle of Berlin was lost. In the last year of the war, after we had successfully invaded France and the Mariana Islands, our bombers were finally able to destroy the cities of Germany and Japan, but the decisive battles happened earlier and were fought by armies and navies. "Victory Through Air Power" turned out to be an illusion. If we had been better prepared to fight with armies and navies, we would probably have won the war sooner.

After this summary of world history dominated by illusions of military glory, Weinberg comes to the politics of the present day. He finds it still dominated by military illusions. Here I particularly recommend his chapters on political themes: "The Growing Nuclear Danger," "Ambling Toward Apocalypse," and "The Wrong Stuff." Since 1945 the dominating illusion has been nuclear weapons. The possession of nuclear weapons now gives people and governments an illusion of power, like the illusions of knights on horseback in the Middle Ages, dreadnoughts in 1905, and strategic bombing in 1936. The United States is repeatedly engaged in costly wars that drag on inconclusively, and we have never found a way to use nuclear weapons effectively. It seems that nuclear weapons cannot be used for any sane military purpose. They are effective for destroying cities and for killing large numbers of people indiscriminately, and for nothing else.

According to Weinberg, the existence of tens of thousands of nuclear weapons is the chief danger to United States security in the modern world. So long as these weapons exist, technical accidents and political miscalculations can easily escalate into an orgy of nuclear killing that would destroy our civilization. No other catastrophe could be as nearly total and as permanent as this one. To get rid of these weapons is more important to our survival than any other objective of our foreign policy.

Second to nuclear weapons, the next most dangerous military illusion is missile defense, a system of defensive rockets or other high-technology weaponry capable of shooting down incoming nuclear missiles. To many people who understand the danger and the uselessness of nuclear weapons, missile defense seems to be a preferable alternative. If a strategy based on nuclear weapons is unsafe and immoral, then a strategy based on active defense against nuclear weapons might be better for our safety as well as for our morality. For forty years, the United States made fruitless efforts to develop a defense against nuclear missiles, and for forty years Weinberg vigorously opposed these efforts. He argues that national security based on missile defense is even more illusory than security based on nuclear weapons. Missile defense can be defeated by concealing real nuclear warheads in a swarm of cheap decoys that look like warheads. In the arms race between offensive missiles and missile defense, the defense can never win because the defender can never know in advance what the offense will do.

The illusion of missile defense decreases our security because it encourages every government possessing nuclear weapons to accumulate larger numbers to overcome possible defenses. I have argued in the past that missile defense might actually be helpful if it were part of a totally defensive strategy, with defenses giving everyone some reassurance of safety as offensive nuclear forces are reduced to zero. I agree with Weinberg that our present efforts to develop missile defense, while also maintaining an offensive nuclear strategy, are worse than useless. Missile defense conforms to the historical pattern of military illusions, combining technical failure with political folly.

A third political folly that Weinberg attacks is the manned exploration of space. He looks at manned space missions in the context of the history of astronomy. Astronomy is the oldest science. For 2,500 years, astronomy led mankind to a true understanding of the way the universe works. From the beginning, instruments were the key to understanding. The first astronomical instrument was the gnomon, a simple vertical post whose shadow allowed the Babylonians and the Greeks to measure time and angle with some precision. The legacy of Babylonian mathematics still survives in the sixtyfold ratios of our units of time: hours and minutes and seconds. After the gnomon came the sundial, the telescope, the chronometer, the computer, and the spacecraft.

Now we are living in a golden age of astronomy, when for the first time our instruments give us a clear view of the entire universe, out in space to the remotest galaxies, back in time all the way to the beginning. Our instruments are telescopes on tops of mountains and on spacecraft in orbit, increasing their capabilities by leaps and bounds as our data-handling skills improve. It takes us only about ten years to build a new generation of instruments that give us radically sharper and deeper views of everything in the sky.

Weinberg contrasts this ongoing triumph of scientific instruments with the abject failure of the American program of manned missions. Our unmanned missions to explore the planets and stars and galaxies have made us truly at home in the universe, while our manned missions after the Apollo program to land on the moon have been scientifically fruitless. Forty years after Apollo, the manned program is still stuck aimlessly in low orbit around the earth, while politicians debate what it should try to do next.

American scientists mostly share Weinberg's view of space. They see the unmanned exploration of space as a success and the manned exploration as a failure. I was lucky to be exposed to a different view in March 2009 when I watched a Russian space launch. The launch was a public ceremony in which the whole community participated. In Russia you do not go into space to do science. You go into space because it is a part of human destiny. To be a cosmonaut is a vocation rather than a profession. Konstantin Tsiolkovsky, the schoolteacher who worked out the mathematics of interplanetary rocketry in the nineteenth century, said, "The earth is the cradle of the mind, but we cannot live forever in a cradle." It may take us a few centuries to get to the planets, but we are on our way. We will keep going, no matter how long it takes.

The American space culture as Weinberg articulates it is only half of the truth. The Russian space culture is the other half. If you think as Americans do, on a time scale of decades, then unmanned missions succeed and manned missions fail. The grandest unmanned missions, such as the Cassini mission now exploring the satellites of Saturn, take about one decade to build and another decade to fly. The grandest manned mission, the Apollo moon landing, ended after a decade and could not be sustained. The time scale of a decade is fundamentally right for unmanned missions and wrong for manned missions. If you think as Russians do, on a time scale of centuries, then the situation is reversed. Russian space science activities have failed to achieve much because they did not concentrate their attention on immediate scientific objectives. Russian manned mission activities, driven not by science but by a belief in human destiny, keep moving quietly forward. There is room for both cultures in our future. Space is big enough for both.

After history and politics, Weinberg's other main theme is science. He writes about science with the sensitivity of a teacher who has taught courses on Physics for Poets, emphasizing basic ideas and avoiding technical details. In this book there are no equations. Chapters with titles such as "Can Science Explain Everything? Anything?" and "The Future of Science, and the Universe" describe Weinberg's philosophical conclusions and say little about the tools of his trade.

His philosophy includes a streak of militant atheism. He has a strong aversion to all religious beliefs and a particularly intense dislike for organized religions such as Christianity and Islam. He quotes with approval the mathematician Paul Erdös, who never used the word "God" but instead spoke of the "Supreme Fascist." He agrees with the statement of Thomas Jefferson: "The Christian God is a being of a terrific character—cruel, vindictive, capricious, and unjust." After this eloquent condemnation of Christianity, Weinberg's chapter "A Deadly Certitude" ends with an even sharper condemnation of Islam. The chapter is a review of Richard Dawkins's book *The God Delusion* (2006). Weinberg concludes:

> Dawkins treats Islam as just another deplorable religion, but there is a difference. The difference lies in the extent to which religious certitude lingers on in the Muslim world, and in the harm it does....I share Dawkins's lack of respect for all religions, but in our times it is folly to disrespect them all equally.

I find it ironic that Weinberg, after declaring so vehemently his hostility to religious beliefs, emerges in his writing about science as a man of faith. He believes passionately in the possibility of a final theory. He wrote a book with the title *Dreams of a Final Theory** in which the notion of a final theory permeates his thinking. A final theory means a set of mathematical rules that describe with complete generality and complete precision the way the physical universe behaves. Complete generality means that the rules are obeyed everywhere and at all times. Complete precision means that any discrepancies between the rules and the results of experimental measurements will be due to the limited accuracy of the measurements.

For Weinberg, the final theory is not merely a dream to inspire his brilliant work as a mathematical physicist exploring the universe. For him it is an already existing reality that we humans will soon discover. It is a real presence, hidden in the motions of atoms and galaxies, waiting for us to find it. The faith that a final theory exists, ruling over the operations of nature, strongly influences his thinking about history and ethics as well as his thinking about science.

I have profound respect for Weinberg as a scientist. I also have profound respect for his faith, although I do not share it. I accept the possibility that he may be right and I may be wrong. I do not forget the disagreement we had forty-four years ago about a hypothetical particle called the W. In chapter 21 I tell the story of this disagreement. The letter W does not stand for Weinberg, but it was Weinberg who imagined it before it was discovered. Weinberg believed that the W particle must exist, because he needed it as an essential component of the theory with which he unified the weak forces in nature. I believed that the W particle could not exist because its existence would contradict a mathematical argument that I held dear. His belief was based on physical intuition, mine on mathematical calculation. A few years later the W particle was directly observed. My mathematical

*Pantheon, 1992.

argument turned out to be irrelevant. I was happy to celebrate Weinberg's triumph, and consoled myself with a quotation from my favorite poet, William Blake: "To be an Error and to be Cast Out is a Part of God's Design." Blake had another piece of wisdom, "Opposition is true Friendship," which made it easy for us to remain friends. As members of the scientific community, we can disagree passionately about facts and theories and still be friends.

Since Weinberg was right about the W particle, why do I not believe that he is right about the final theory? I distrust his judgment about philosophical questions because I think he overrates the capacity of the human mind to comprehend the totality of nature. He has spent his professional life within the discipline of mathematical physics, a narrow area of science that has been uniquely successful. In this narrow area, our theories describe a small part of nature with astonishing clarity. Our ape brains and tool-making hands were marvelously effective for solving a limited class of puzzles. Weinberg expects the same brains and hands to illuminate far broader areas of nature with the same clarity. I would be disappointed if nature could be so easily tamed. I find the idea of a final theory repugnant because it diminishes both the richness of nature and the richness of human destiny. I prefer to live in a universe full of inexhaustible mysteries and to belong to a species destined for inexhaustible intellectual growth.

Isaac Newton, the scientist who took the biggest single step toward the understanding of nature, saw clearly how far he was from any final theory. "I do not know what I may appear to the World," he wrote toward the end of his long life, "but to myself I seem to have been only like a boy playing on the seashore, and diverting myself in now and then finding a smoother pebble or a prettier shell than ordinary, whilst the great ocean of truth lay all undiscovered before me."

Newton wrote more modestly than Weinberg of the ability of the human mind to penetrate the mysteries of nature. Newton was a devout Christian, as dedicated to theology as he was to science. Newton was no fool.

Note added in 2014: Weinberg wrote in response to the review:

> You may be right about prospects for a final theory. My "faith" is really not much more than a judgment that it's worth trying. Your Newton quote is apposite, but on the other hand, it was Newton who also said, "I wish we could derive the rest of the phenomena of nature by the same kind of reasoning as for mechanical principles. For I am induced by many reasons to suspect that they may all depend on certain forces."

11

SILENT QUANTUM GENIUS

WHY SHOULD ANYONE who is not a physicist be interested in Paul Dirac? Dirac is interesting for the same reasons that Albert Einstein is interesting. Both made profound discoveries that changed our way of thinking. And both were unique human beings with strong opinions and strong passions. Besides these two major similarities, many details of their lives were curiously alike. Both won the Nobel Prize in Physics, Einstein in 1921 and Dirac in 1933. Both had two children of their own and two stepchildren from a wife's previous marriage. Both were intensely involved in the community of professional scientists in Europe when they were young. Both of them emigrated to the United States and became isolated from the American scientific community when they were old. The main difference between them is the fact that Einstein was one of the most famous people in the world while Dirac remained obscure.

There are many reasons why Einstein became inordinately famous. The main reason is that he enjoyed being famous and entertained the public with provocative statements that made good newspaper headlines. Dirac had neither the desire nor the gift for publicity. He discouraged inquisitive journalists by remaining silent. Einstein has had dozens of books written about him, while Dirac has only two, *Dirac: A Scientific Biography* by Helge Kragh, published in 1990, and *The Strangest Man: The Hidden Life of Paul Dirac, Mystic of the Atom*, a new biography by Graham Farmelo.* The Kragh biography is full of equations and is addressed to experts only. The enormous fame of Einstein and the obscurity of Dirac have given the public a false picture of the two revolutions that they led. The public is aware of one revolution and correctly gives credit for it to Einstein. That was the revolution that changed the way we think about space and time. The new way of thinking was called relativity.

The second revolution that came ten years later was more profound and changed the way we think about almost everything, not only in physics but in chemistry and biology and philosophy. It changed the way we think about the nature of science, about cause and effect, about past and future, about facts and probabilities. This new way of thinking was called quantum mechanics. The second revolution was led by a group of half a dozen people, including Einstein. It does not belong to a single leader. But the purest and boldest thinker of the second revolution was Dirac. If we wish to give the second revolution a human face, the most appropriate face is Dirac. Farmelo writes that "in one of his greatest achievements," Dirac arranged

> what had seemed an unlikely marriage—between quantum mechanics and Einstein's theory of relativity—in the form of an exquisitely beautiful equation to describe the electron. Soon afterwards, with no experimental clues to prompt him, he used his equation to predict the existence of antimatter, previously unknown particles with the same mass as the corresponding particles of matter but with the opposite charge. The success of

*Basic Books, 2009.

> this prediction is, by wide agreement, one of [the] most outstanding triumphs of theoretical physics.

In Farmelo's book we see Dirac as a character in a human drama, carrying his full share of tragedy as well as triumph. He is as strange a figure as Einstein. He is less famous because he preferred to fight his battles alone.

The title, *The Strangest Man: The Hidden Life of Paul Dirac, Mystic of the Atom*, is not well chosen. The British edition published by Faber and Faber has a better title, with "Mystic of the Atom" replaced by "Quantum Genius." Mystic and genius are not synonymous. The phrase "The Strangest Man" is a quote from Niels Bohr, the great Danish physicist who invited Dirac to visit his Institute for Theoretical Physics in Copenhagen in 1926 when Dirac was twenty-four years old. Bohr said many years later that Dirac was the strangest man who had ever come to his institute.

It was true that Dirac's inner life was well hidden. He did not like to reveal what he was thinking, either about science or about himself. But Bohr did not say that Dirac was a mystic, and it is not true. Dirac was the opposite of a mystic. He worked in a straightforward way, trying out mathematical schemes to describe the way nature behaves. What was strange about Dirac was not mysticism but formidable concentration of attention upon a single problem. He was silent and aloof because he liked to think about one thing at a time. In his choice of problems to think about, he was guided by his ability to set aside irrelevancies, to see clearly what was important and what was not. For him, most of the subjects that people talk about in everyday conversation were irrelevant, and so his conversation was mostly silence.

Although Dirac spoke little about himself, he preserved almost all the letters and papers that he received from his family and friends, all the way back to his childhood. These papers are now accessible in the Dirac archive at Florida State University. They provide a solid documentary basis for Farmelo's biography. In spite of Dirac's legendary silence, we know more about his early life than we know about his more talkative contemporaries. Farmelo also interviewed everyone still alive who had known Dirac, and obtained detailed accounts of rare conversations in which Dirac as an old man talked at length about his youth. The most dramatic of these conversations was reported by Kurt Hofer, a biologist colleague of Dirac at Florida State University. Farmelo places Hofer's story at the beginning of his biography, emphasizing its importance for the understanding of Dirac's struggles. The truth of the story is confirmed by other witnesses and documents.

Paul Dirac's father was Charles Dirac, a Swiss schoolteacher who taught modern languages in the English city of Bristol. He was a capable but harsh teacher. Paul's mother, Florence, was twelve years younger and dominated by her husband. Paul had an older brother, Felix, and a younger sister, Betty. According to Hofer, Charles made Paul's life miserable by insisting that he speak only French at home and punishing him when he made grammatical mistakes. Since talking brought punishment, Paul acquired the habit of silence. Charles was habitually unfaithful to Florence, and the pair were barely on speaking terms. Paul was close to his mother, Betty to her father, and Felix to neither. Felix suffered acutely from comparisons with his brilliant brother. When Paul was twenty-three and Felix twenty-five, Felix killed himself. By that time Paul had escaped from the hate-filled home and enrolled with a scholarship at St. John's College, Cambridge. Paul's hatred of his father endured until he talked with Hofer more than fifty years later. Paul said, "I never knew love or affection when I was a child," and speaking of his father, "I owe him absolutely nothing."

In spite of these inner torments, Dirac had a remarkable talent for friendship. His closest friend at Cambridge was Peter Kapitza, the charismatic Russian experimental physicist who later won a Nobel Prize for discovering the superfluidity of liquid helium. Kapitza then lived and worked in England but spent his summers in Russia. Dirac several times went to Russia for long holidays with Kapitza and other Russian friends, climbing mountains and enjoying the comforts of Kapitza's country club in the Crimea. Dirac also tried his hand at experimental work in Kapitza's laboratory.

In 1934 Stalin decided to keep Kapitza in Russia and forbade him to return to England. Dirac then traveled to Russia in an unsuccessful campaign to get Kapitza out. Kapitza was depressed, and Dirac stayed for several weeks at his dacha to restore his spirits. In the end, a deal was arranged so that Kapitza stayed in Russia and the Soviet government paid for all of his experimental equipment to be shipped from Cambridge to Moscow. He could then continue his experimental work in Russia as the director of his own institute. Dirac and Kapitza remained friends at a distance and enjoyed a happy reunion in Cambridge thirty years later. Kapitza was a great talker and Dirac was a great listener, so they were well matched.

Dirac was also a faithful friend to several of the other pioneers of quantum mechanics, particularly Niels Bohr and Werner Heisenberg. Heisenberg stayed in Germany through World War II. Although he disliked Hitler, he was a patriotic German and considered it his duty to serve his country and share its fate. After the war, he suffered grievously from the hostility of former friends who never forgave him for leading the German Uranium Project, an abortive effort that never came close to developing a nuclear bomb. Dirac went out of his way to be friendly to him, saying that Heisenberg had behaved reasonably in an extremely difficult situation. Dirac had seen his friends in Russia forced to make difficult choices under a capricious government and understood the pressures under which they lived. He said, "It is easy to be a hero in a democracy."

Dirac's talent for human relationships was shown most spectacularly in his marriage to Manci Balázs, which lasted for forty-seven years until his death. Manci was a temperamental Hungarian widow, accustomed to an aristocratic lifestyle. Dirac was a quiet fellow who enjoyed the company of outgoing people. With their opposite qualities they were well matched, just as Dirac and Kapitza had been. Manci took care of Dirac and organized his social life. He enjoyed his stepchildren and his children, letting them run free and be themselves, avoiding the harsh treatment that had alienated him from his own father. He enjoyed working long hours in his garden, growing flowers in peacetime and vegetables in wartime. He enjoyed listening quietly while Manci and her friends talked.

Dirac and Manci lived together in Cambridge for thirty-four years, most of the time amicably. Their most serious disagreements resulted from the fact that Dirac loved the quiet routine of Cambridge while Manci found it boring. According to legend, there was a time when Manci became infuriated as she was serving dinner and said to Dirac, "What would you do if I left you?" After an interval of silence, Dirac replied calmly, "I'd say 'Goodbye dear.'"

I enjoyed a firsthand glimpse of Dirac when I came to listen to his lectures as a seventeen-year-old undergraduate at Cambridge. Like Bohr in Copenhagen fifteen years earlier, I found him strange. That was in 1941, for England the third year of World War II. Because of the war the number of students was small, but Dirac lectured faithfully to his little band of listeners every Monday, Wednesday, and Friday morning. His lectures were mostly a verbatim recitation of his book *The Principles of Quantum Mechanics*. In the introduction to that book he says:

> This has necessitated a complete break from the historical line of development, but this break is an advantage through enabling the approach to the new ideas to be made as direct as possible.

In other words, he gives us the abstract mathematical scheme that describes how nature behaves, without explaining the earlier history of physical ideas out of which this scheme arose. In those days I wrote a letter to my parents once a week, so I have a contemporary record of the impact of his lectures. I wrote in February 1942:

> Dirac reached a climax of difficulty in his last lecture. [Another professor] has promised to invite me next term to a tea-party with Dirac. Dirac is a man I should like to talk to,...but he lectures like a gramophone record and nobody seems to know him at all.

There is no record of the tea party with Dirac, if it ever happened. But nine months later there was another tea party, which I recorded on November 30, 1942:

> There were present also two young Diracs, by name Gabriel and Judith. Gabriel is 17 and a first-year undergraduate from St. John's, Judith is 15 and at a school in Cambridge. They are Hungarians by upbringing, and know a lot about Central Europe; also they used to row on the Danube with von Neumann the great topologist, an almost legendary figure now in America. Gabriel is a very ardent member of the communist party, and thus kept the ball rolling from the start. The young Dirac is reading mathematics but is at present more interested in politics.

Gabriel and Judith were the stepchildren who moved into Dirac's life when he married their mother in 1937. As teenagers they were outstandingly bright and lively. When I met them at the tea party, the Battle of Stalingrad was raging, the Soviet Union was our gallant ally bearing the main brunt of the war against Hitler, and for a bright teenager to be a Communist was not unusual. Dirac himself was not a Communist, but he was strongly socialist. He had traveled many times to Russia, where some of his closest friends were living, and he had been welcomed by the Soviet government as a visitor even when his friends were in political trouble.

Six years later, when I arrived at the Institute for Advanced Study in Princeton for the first time, I encountered Dirac with his own daughters, Mary and Monica, then aged eight and six. Here was the scene, dated September 14, 1948:

> When I visited the Institute there were more children there than grown-ups, Dirac with his family shortly leaving for England, and various other children playing cowboys and Indians, and Von Neumann looking rather vague in the midst of the confusion.

This vivid scene, with Mary and Monica chasing each other around the table in the middle of the institute common room, was very different from the staid formality of college common rooms in Cambridge to which both Dirac and I were accustomed.

When Dirac approached the age of retirement from his professorship in Cambridge, Manci decided that the time had come to move him to America. Mary and Monica were both married and settled there, and Manci had had enough of England. Manci took Dirac to Tallahassee, where Mary was living, and the Physics Department of Florida State University offered him a job as "visiting eminent professor." Dirac accepted the offer, and lived the last thirteen years of his life in Tallahassee as an honored sage, becoming more sociable and more talkative as he grew older. Manci was happy to open her house to a stream of American friends who came to escape the northern winters and to talk with her famous husband. Manci and Dirac understood each other's needs and achieved a certain serenity in their old age.

Farmelo ends his book with two provocative chapters, one entitled "On Dirac's Brain and Persona" and the other "Legacy." Each of them raises intriguing questions. The questions are perhaps unanswerable. Farmelo gives them tentative answers based on his own opinions. I shall explain why I give them different answers, equally tentative, based on my opinions. The chapter on Dirac's brain and persona asks whether he was autistic. Autism was until recently a rare disease, characterized by mental disorders that made the patient incapable of a normal life. The main symptom was a failure to achieve or understand social relationships with other human beings. This was often accompanied by failure to speak or by severe retardation of speech. The typical autistic child was obsessed with repetitive activities, resistant to any change in established routines, and uninterested in communication with family or friends.

Judged by these criteria, Dirac was clearly not autistic. My wife found him a friendly and amusing companion when she went for a walk with him in Princeton. He was intensely and personally involved with his physicist friends such as Kapitza, Heisenberg, and Bohr. He had close friendships with at least three women before he married Manci. And he had normal fatherly relationships with his stepchildren and children. If Dirac was autistic, then the word "autism" must have a different meaning.

During the last twenty years the concept of autism has been broadened so as to include a much wider variety of people. Debates have raged among the experts concerning the criteria for a medical diagnosis. The term "autistic spectrum disorder" has been introduced to give official recognition to the wider meaning of autism. As a result, autism is now no longer rare, and it includes many people who can function normally in society but have the classic symptoms of autism in milder form.

To be autistic in the wider sense, it is enough to be insensitive to other people's feelings, to be more interested in things than in people, and to be intensely interested in things that normal people find bizarre. The diagnosis of autistic spectrum disorder includes a wide spectrum of disability, from mute and severely retarded people languishing in mental institutions to articulate and gifted university professors living active professional lives. Temple Grandin is a famous example of the high-functional autistic: a professor of engineering, the author of several books, and a world-class expert on the design of buildings and machines for the humane handling of livestock. Farmelo describes two aspects of Grandin's autism—her sensitivity to sudden sounds and the visual nature of her thought processes—and makes a case for Dirac's autism by analogy.

"Asperger's syndrome" is an alternative name for the high-functional end of the autistic spectrum. Hans Asperger was an Austrian psychologist who studied children who were socially inept but intellectually sharp. Several parents of my acquaintance are proud to claim that their gifted children "have a touch of Asperger" when the children develop a passion for painting or mathematics. Asperger's syndrome has become a distinction rather than a disease. If every child who is silent and withdrawn and has an unusual talent has a touch of Asperger, then Dirac certainly had a touch of Asperger. If Asperger's syndrome is included in the autistic spectrum, then Farmelo is justified in concluding that Dirac was autistic.

The definition of autism is today based on symptoms that are poorly defined and on medical judgments that are largely subjective. That is why the posthumous diagnosis of Dirac as autistic or nonautistic is a matter of opinion. But in the future, an objective diagnosis may become possible. We have strong evidence that autism is associated with anatomical abnormalities in the brain, and that autism is heritable. When this evidence from neurology and genetics has been consolidated, it is possible that the diagnosis of autism based on symptoms will be replaced by a diagnosis based on the objective observation of brains and genomes. At some time in the future, when the intricacies of brains and genomes are understood in detail, a reliable posthumous diagnosis of Dirac's personality may become possible, provided that some fragment of tissue carrying his DNA has been preserved.

Since the era of diagnosis based on DNA has not yet arrived, I base my tentative diagnosis on anecdotal evidence. Two anecdotes in Farmelo's book strike me as strong evidence that Dirac's peculiarities had nothing to do with autism. Both episodes occurred before he married Manci and became domesticated. In 1935, when Kapitza and his wife were detained in Russia, their two sons were left behind in England, and the Kapitzas appointed Dirac as legal guardian of the boys. Dirac took care of the boys until they joined their parents in Russia. While the boys were with him, Guy Fawkes Day occurred, the English equivalent of July 4, traditionally celebrated with bonfires and fireworks. Dirac organized a fireworks show for the boys.

The second anecdote concerns another famous Russian physicist, George Gamow, who was also a close friend of Dirac. Gamow was notorious as a practical joker. He had emigrated from Russia and settled in Washington with his wife. Dirac was traveling in Florida and saw some alligators for sale. He decided to give Gamow a taste of his own medicine, bought a baby alligator, and mailed it anonymously in a parcel to Washington. The joke succeeded even better than Dirac intended. Gamow's wife opened the parcel, was seriously bitten by the alligator, and accused her husband of perpetrating the joke. Dirac had hit two birds with one stone. He let a month pass before confessing that he was the guilty party. These two stories show us Dirac as he was in his thirties, a young man with a fondness for children and a robust sense of humor, very far from the pathological self-absorption that is the basic symptom of autism.

The last chapter of Farmelo's book concerns Dirac's legacy to later generations. The legacy consists of three parts: first, the laws of nature that Dirac discovered in his wonder years from 1925 to 1933; second, the doctrine of mathematical beauty that he preached for the remaining fifty years of his life; and third, his distaste for philosophical interpretation of his discoveries. For practicing scientists, the chief legacy of Dirac is the starburst of discoveries that he made as a young man. His legacy to nonscientists is not so clear. Farmelo emphasizes the doctrine that he preached later, proclaiming that mathematical beauty is the key to scientific truth. In order to discover the true laws of nature, the searcher after truth should pay more attention to abstract beauty than to practical details. The beauty and simplicity of the laws of nature would be revealed in abstract mathematics. The second legacy is summarized in a statement that Dirac wrote at the end of his life: "If you are receptive and humble, mathematics will lead you by the hand."

The doctrine of mathematical beauty is itself beautiful, and there is no doubt that Dirac believed it to be true. But it does not agree well with the historical facts. During the wonder years when he was making his great discoveries, his thinking was more concerned with practical details and less with abstract beauty. And during the long second half of Dirac's life, when he was preaching the doctrine of mathematical beauty, it did not lead him to important new discoveries.

During Dirac's middle years, the grand edifice of modern particle physics was growing up around him, with discoveries of new particles and new symmetries bursting out in rapid succession. Nature was screaming at him to pay attention to her revelations. But his love for abstract beauty told him to stay aloof. He ignored the new discoveries of particles and symmetries, because he judged them to be too complicated, not beautiful enough to be true. Instead of listening to Nature, he was telling Nature how to behave. As a result, the second half of his life was comparatively sterile.

In addition to the two legacies of discoveries and aesthetic principles, Dirac also left a third legacy, which I consider precious but Farmelo does not. This legacy is Dirac's refusal to engage in philosophical arguments about the interpretation of quantum mechanics. These philosophical debates raged during his lifetime and have raged even more fiercely after his death. Dirac took no part in these debates and considered them to be meaningless. He said, as Galileo said three hundred years earlier, that mathematics is the language that nature speaks. When expressed in mathematical equations, the laws of quantum mechanics are clear and unambiguous. Confusion arises from misguided attempts to translate the laws from mathematics to human language.

Human language describes the world of everyday life and lacks the concepts that could describe quantum processes accurately. Dirac said we should stop arguing about words, stay with mathematics, and allow the philosophical fog to blow away. I consider Dirac's disengagement from verbal disputes about the meaning of quantum mechanics to be an essential part of his legacy. But I am, as usual, in the minority.

Note added in 2014: In response to this review, I received a number of illuminating letters from friends and relatives of the Dirac family. I encouraged the authors to send copies of their letters to the Dirac archive, which is maintained at Florida State University at Tallahassee.

12

THE CASE FOR FAR-OUT POSSIBILITIES

SINCE PEOPLE BEGAN to wonder about human destiny, there have always been prophets of hope and prophets of doom. Long ago in Mesopotamia, as recorded in the book of Genesis, Abraham fell on his face and God talked with him, saying:

> Behold, my covenant is with thee, and thou shalt be a father of many nations.... And I will make thee exceeding fruitful, and I will make nations of thee, and kings shall come out of thee. And I will establish my covenant between me and thee and thy seed after thee in their generations for an everlasting covenant.

Abraham was the first prophet of hope in the Western tradition. He set the pattern of our culture. He was a traveler, moving into a new country to take possession of it for his descendants. A little later, other prophets of hope, Gautama Buddha and Lao Tse, started other traditions in other places. Meanwhile, in the West, Jeremiah the prophet of doom raised his voice in Jerusalem against Abraham:

> The word of the Lord came also unto me, saying, thou shalt not take thee a wife, neither shalt thou have sons or daughters in this

> place. For thus saith the Lord concerning the sons and concerning the daughters that are born in this place, and concerning their mothers that bare them, and concerning their fathers that begat them in this land. They shall die of grievous deaths; they shall not be lamented; neither shall they be buried; but they shall be as dung upon the face of the earth: and they shall be consumed by the sword, and by famine; and their carcasses shall be meat for the fowls of heaven, and for the beasts of the earth.

Other prophets of doom proclaimed in other traditions the anger of gods and the helplessness of humans. The dialogue between Abraham and Jeremiah continues today. It is still one of the main themes of our history. So what is new?

One thing that is new is modern science. Science has not displaced religion as the way most people approach the problems of our destiny, but science allows us all to look at these problems in a new way. Francis Bacon, the major prophet of modern science in the British tradition, did not proclaim the word of the Lord but spoke with his own more modest voice: "If we begin with certainties, we will end in doubt, but if we begin with doubts and bear them patiently, we may end in certainty."

Bacon was writing at the beginning of the seventeenth century, when religious wars were raging in Europe, when Pilgrim fathers filled with Abrahamic hopes were building a new world in America, when Puritan divines filled with visions of jeremiad doom were preaching hellfire and damnation. He offered a third alternative to the certainties of heaven and hell: the alternative of patient inquiry. He told us to ask questions instead of proclaiming answers, to collect evidence instead of rushing to judgment, to listen to the voice of nature rather than to the voice of ancient wisdom. Bacon predicted accurately the growth of modern science. In the centuries since he wrote, modern science has transformed the problem of human destiny. Destiny is now no longer an unalterable fate, irreversibly good or evil. Destiny has become a continuing experiment in which we are free to learn from our mistakes.

In the modern world of life and death to which we all belong, a crucial problem of our destiny is the size of human populations. It seems to be a simple problem. How many people should there be? How many babies should we raise? But modern science has twice transformed the nature of the problem. In the eighteenth century, science started industrial and medical revolutions that caused a rapid growth of populations. In the twentieth century, science started social revolutions that caused an equally rapid decline of birth rates.

It was easy to understand, as Robert Malthus pointed out in his famous *Essay on the Principle of Population* in 1798, how new technology had caused population growth that might in turn cause a disproportionate growth of human misery. It is more difficult to understand, in the world of today, how birth rates fell rapidly in large parts of the world while remaining high in others. It appears that birth rates fell sharply for different reasons in different places, in China because of draconian rules imposed by the government, in Europe and America because a large fraction of women became educated and economically independent. Meanwhile, birth rates remain high in Africa and in parts of Asia where societies are male-dominated and women are mostly illiterate.

Three facts emerge clearly from the history of the last three centuries. First, a huge experiment is in progress, exploring various ways of dealing with the problem of population. Second, no central authority is in charge. Third, the result of the experiment is still in doubt. Many kinds of disaster resulting from population explosion or population collapse are still possible. Nevertheless, the results of the experiment up to the present time are encouraging. The disasters predicted by Malthus did not occur. Populations in several parts of the world with different political and ethical traditions were successfully controlled by different methods. It appears to be generally true that rising wealth and communication and education in any society produce a rapid fall in birth rates. This is an experimental conclusion, subject to criticism and correction. It does not tell us that the problem of population is finally solved. It tells us that the solution of the problem is still in our hands, to be explored by continued experiment and by correction of errors.

When we look to the remote future, the main problem of our destiny will not be the size of populations but their quality. Shall we remain a single species bound together by bonds of family and kinship, or shall we evolve into many diverse species as our vertebrate ancestors did in the past? Either alternative brings losses as well as gains. If we remain single, we lose vast opportunities to explore new ways of living and thinking. We lose the historic power of biological evolution to try out new experiments and to create new designs of body and mind. If we diversify, we lose the brotherhood of man. We lose the shared loyalties and traditions that made us what we are. By separating into alien species, we open endless possibilities of future strife and irreconcilable quarrels.

Perhaps these dangers can be mitigated if the diversification of human nature is combined with an expansion of our habitat from one planet to a multitude of communities spread out over the universe. Those of us who choose to stay on this planet should remain brothers, while those who choose to experiment with new creative possibilities should move far enough away that the failure of their experiments will not endanger those who stayed at home. The vastness of the universe allows us to dream of an infinite future for humanity, with bodies and minds spreading in space and expanding in quality far beyond anything that we can imagine. That is why David

Deutsch gave his book the title *The Beginning of Infinity*.* The subtitle, "Explanations That Transform the World," carries the central message of the book. It says that our destiny is to be explainers of the world around us, and explaining is the key to mastery.

Deutsch has an important message. He writes clearly and thinks wisely. His book could help to push the world toward better ways of dealing with its problems. It is written for concerned citizens and not only for philosophers. I hope many concerned citizens will read it and take its message to heart. Unfortunately, Deutsch is himself a philosopher, with a fondness for abstruse philosophical arguments. Fortunately, he puts his plain language and his abstruse philosophizing into separate chapters. The common reader should skip the technical chapters and pay attention to the others. The difficult chapters 11 and 12, "The Multiverse" and "A Physicist's History of Bad Philosophy," ought to have been published as a separate book, addressed to a different audience. They have little connection with the outstandingly lucid chapters 10 and 13, "A Dream of Socrates" and "Choices," which stand immediately before and after them. The difficult chapters are for readers who share Deutsch's view of the nature and purpose of philosophy.

Philosophy can be regarded as a branch of science or as a branch of literature. For Deutsch, philosophy is a collection of explanations and arguments that are either right or wrong. For me, philosophy is a collection of stories. For Deutsch, the only philosopher who deserves unconditional respect is Karl Popper, because he alone asked the right questions and gave them the right answers. For me, the great philosophers are those like Plato and Bertrand Russell who happen to be good writers. In one of his lighter moments Russell expressed a view of philosophy similar to mine: "Science is organized common

*Viking, 2011.

sense; philosophy is organized piffle." In my view, Deutsch becomes a true philosopher when he forgets his technical arguments and tells evocative stories.

Deutsch sums up human destiny in two statements that he displays as inscriptions carved in stone: "problems are inevitable" and "problems are soluble." His chapter "The Spark" introduces these statements and explains their meaning. They apply to all aspects of human activity, to ethics and law and religion as well as to art and science. In every area, from pure mathematics and logic to war and peace, there are no final solutions and no final impossibilities. He identifies the spark of insight, which gave us a clear view of our infinite future, with the beginning of the British Enlightenment in the seventeenth century. He makes a sharp distinction between the British Enlightenment and the Continental Enlightenment, which arose at the same time in France.

Both enlightenments began with the insight that problems are soluble. Both of them engaged the most brilliant minds of that age in the solution of practical problems. They diverged because many thinkers of the Continental Enlightenment believed that problems could be finally solved by utopian revolutions, while the British believed that problems were inevitable. According to Deutsch, Bacon transformed the world when he took the long view, foreseeing an infinite process of problem-solving guided by unpredictable successes and failures. Deutsch's version of history is narrow. It is Whig history, portraying human destiny as a triumph of parochial British ideas and institutions.

Long before Bacon, thinkers in China were taking a long view of history and pushing it along a different path, and Socrates in Greece was teaching us to search for wisdom by asking questions rather than by knowing the answers. Many diverse cultures were converging to the conclusion that humans have a choice. If we want to, we can be the spark, transforming the universe from a purposeless machine into a creative community of living creatures always asking new questions and struggling to find new answers.

Returning to present-day problems in his chapter "Choices," Deutsch discusses politics and economics. Two questions have dominated the study of political economy in the past. How should we choose our rulers? And how unequally should we distribute wealth between rich and poor? In recent debates over the choice of rulers, people have usually asked the wrong question. They asked, who are the best rulers? They assumed that if this question were answered, then we should allow the best rulers to rule and the problem of good government would be solved.

But history taught us long ago that this is the wrong question. There are no best rulers, because power corrupts and circumstances change. Rulers often begin well and then make stupid mistakes. The English civil war between King Charles I and his Parliament demonstrated clearly that the concept of a best ruler was an illusion. The king, claiming to rule by divine right, abused his power flagrantly, and the parliamentary leaders rose in rebellion against him. The Parliament won the war, beheaded the king, and appointed Oliver Cromwell to rule in his place. When Cromwell died, the Parliament decided after all to invite the son of the murdered king back.

Chastened by defeat and exile, the new king ruled more gently and more wisely than either his father or Cromwell. During that century, the great debate between monarchy and republic was conducted on an extraordinarily high intellectual level. The two greatest poets of the English language were deeply engaged: William Shakespeare standing close to the monarchs Elizabeth and James, John Milton standing close to Cromwell. Shakespeare's histories and tragedies held up the mirror to kings and queens. Milton's *Areopagitica* held up the mirror to officeholders who try to rule our minds.

The right question to ask was not "Who are the best rulers?" but "How do we make sure that rulers can be peacefully replaced when they rule badly?" Democratic systems of government are designed to answer the latter question. Elections are held not to choose the best rulers but to give us a chance to get rid of the worst without bloodshed. Constitutional monarchy is another solution to the same problem. The present queen of England has no power to rule her country, but she has the power to dissolve Parliament and stop any politician from taking actions that are flagrantly unconstitutional. The perennial problem of government is not to choose the best rulers but to hold bad rulers responsible for their failures.

The division of wealth between rich and poor is a problem similar to the division of political power and has a similar history. The great debate has been between the ideals of ethics and of economics. Social justice demands equality. Fair reward for enterprise and achievement demands inequality. Advocates on both sides of the debate have tended to take extreme positions. Numerous utopian communities have been founded to put egalitarian principles into practice. Few of them have lasted for longer than one generation. Children have a regrettable tendency to rebel against their parents' dreams. Meanwhile, advocates of extreme free-market capitalism have been preaching the gospel of greed. They glorify greed as the driving force that creates new industries and in the end will make everyone wealthy. Unfortunately, in many parts of the world where free-market capitalism prevails, the rich are growing richer and the poor are growing poorer.

The dominant utopian thinker in the great debate over economic power was Karl Marx, who saw the world of the nineteenth century as black and white. Black was capitalism, the existing society of rich factory owners and downtrodden workers, with power concentrated in the hands of the owners. White was communism, the future society of workers seizing power for themselves and owning the means of production. Communism would achieve social justice for the workers after consigning the former owners to the dustbin of history. Marx was a prophet of hope, describing his dreams of the future in language worthy of his Hebrew forerunner Isaiah. "For, behold," wrote Isaiah,

> I create new heavens and a new earth: and the former shall not be remembered, nor come into mind.... The wolf and the lamb shall feed together, and the lion shall eat straw like the bullock: and dust shall be the serpent's meat. They shall not hurt nor destroy in all my holy mountain, saith the Lord.

Looking back on Marx's visions today, we can see that much of what he wrote about capitalism was true and almost everything he wrote about communism was false. As long as he was examining the evidence that he saw around him, he was on firm ground. As soon as he moved from evidence to dogma, his imagination led him wildly astray.

Thanks to the magic of modern data search and rapid communication, I received from a cousin in Australia a copy of the marriage certificate of my great-grandparents Jeremiah and Mary Dyson, married in 1857 in the parish church of Halifax in the industrial north of England, the region where Marx's friend Friedrich Engels had written his classic denunciation of capitalism, *The Condition of the Working Class in England*. Mary did not sign her name. She put her mark X on the certificate. After that, without the help of a Communist revolution, the condition of the Halifax working class slowly improved. They achieved education, and modest prosperity, and the freedom to pursue broader interests. Mary's son became a skilled machine builder, her grandson became a professional musician, and her great-grandson is a scientist.

The gospel according to Marx is a classic example of bad philosophy as defined by Deutsch. Bad philosophers try to improve the human condition by telling the world how to behave. They deceive themselves, imagining that the world will dance to their tune. Good philosophers continue to observe how the world is behaving and try to explain what they observe. Good philosophers improve the human condition by asking questions and correcting errors. The method of good philosophy is to explain and understand how the world behaves, not to prescribe.

The most important improvement of the human condition in the last half-century was the economic transformation of China. If this transformation continues for another half-century and also includes India, more than half of the population of the world will be rich. The way will be open for new and unpredictable transformations to come. China has a long tradition, extending back through thousands of years, of central government organizing large-scale social experiments. Some of the experiments failed and some succeeded. The Chinese tradition encourages the taking of large risks and the ability to recover from calamities. We should hope that the Chinese tradition will continue to be different from ours, so that they will dare to undertake new ventures that our more timid Western rules forbid. It is a pity that Deutsch does not mention China in his book. He ignores half of our heritage. If he had brought China into his vision of the future, his argument for an infinite expansion of human possibilities would have been strengthened.

Of Deutsch's eighteen chapters, the one that I recommend most strongly is "A Dream of Socrates," a lighthearted piece of philosophical fiction. Socrates comes to the oracle at Delphi to ask who is the wisest man in the world. The oracle, speaking for the god Apollo, answers, "No one is wiser than Socrates." Asleep in his hotel the following night, Socrates is visited by Hermes, the messenger of the gods. The two of them enjoy a dialogue of jokes and paradoxes, in which Hermes explains to Socrates all the main points that Deutsch is advocating in his book. The most important point that Hermes explains is that wisdom is achieved by asking questions, that is to say, by following the method that we now call Socratic. After that, Socrates is rudely awakened by young Plato and a bunch of other friends bursting into his room, and Hermes disappears. Socrates tries to explain to Plato what he has learned. Plato scribbles Socrates's words down hastily with a stylus on a writing tablet, but he misunderstands and garbles the message.

Note added in 2014: David Deutsch reappears in chapter 17 as a character in a book by the philosopher Jim Holt. The Deutsch of chapter 17 is a more dogmatic speculator, and I criticize him there more harshly. According to my account, philosophy lost its bite when philosophers moved from the public agora of Athens to the cloistered colleges of Oxford.

13

SCIENCE ON THE RAMPAGE

PHYSICS ON THE FRINGE describes work done by amateurs, people rejected by the academic establishment and rejecting orthodox academic beliefs.* They are often self-taught and ignorant of higher mathematics. Mathematics is the language spoken by the professionals. The amateurs offer an alternative set of visions. Their imagined worlds are concrete rather than abstract, physical rather than mathematical. Many of them belong to the Natural Philosophy Alliance, an informal organization known to its friends as the NPA.

Margaret Wertheim's book discusses her encounters with the natural philosophers. She is interested in them as characters in a human tragedy, with the seriousness and dignity that tragedy imposes. Her leading character is Jim Carter, and her main theme is the story of his life and work. Unlike most of the philosophical dreamers, Carter is a capable engineer and does real experiments to test his ideas. He runs a successful business that gives him leisure to pursue his dreams. He is a man of many talents, with one fatal flaw.

Carter's flaw is his unshakable belief in a theory of the universe

*Margaret Wertheim, *Physics on the Fringe: Smoke Rings, Circlons, and Alternative Theories of Everything* (Walker, 2011).

based on endless hierarchies of circlons. Circlons are mechanical objects of circular shape. The history of the universe is a story of successive generations of circlons arising by processes of reproduction and fission. He verified the behavior of circlons by doing experiments with smoke rings at his home. A smoke ring is a visible manifestation of a circlon. He built an experimental apparatus using garbage cans and rubber sheeting to make long-lived smoke rings under controlled conditions. The fact that smoke rings can interact with one another and maintain a stable existence proves that circlons can do the same. Just as the standard theory of nuclear physics is verified by accelerator experiments, he claims that his theory is verified by his garbage-can experiments at a million times lower cost. Like Coleridge's Ancient Mariner, he tells his story over and over to listeners who will not believe it.

The most dramatic period of Carter's life was the 1970s, when he made a living as a diver collecting abalone from the sea bottom around Catalina Island. His first vision of a circlon was a perfect ring of air bubbles that sometimes rose from the exhaust valve of his underwater breathing apparatus when he exhaled. In those days, abalone were abundant and the demand for them insatiable. He could make enough money in a day of diving to allow him to stay at home for a week and work out the theory of circlons.

The practical limit to his income was the difficulty of transporting large quantities of abalone from the sea bottom to land. He solved this problem by inventing a device called a lift bag, which is a large bag with one compartment for freight and another compartment inflated with air. A small tank of compressed air released into the bag can lift hundreds of times its own weight. The lifting capacity of the bag is the weight of water displaced by the air, and the water weighs a thousand times as much as the air. With the help of his wife, he designed a lift bag that was elegant and user-friendly, using brightly colored materials to improve its underwater visibility. He was soon receiving orders for lift bags from people engaged in underwater operations of all kinds, from raising sunken ships to drilling for oil. The Carter Lift Bag Company was bringing him a larger income than he had ever earned from abalone.

Carter was unaware, until Wertheim told him the news, that his smoke-ring experiments had been done with similar apparatus and for a similar purpose 130 years earlier. William Thomson and Peter Tait, a famous physicist and a famous mathematician, had invented a theory of matter similar to Carter's theory of circlons. They imagined every atom to be a vortex in a hypothetical fluid known as ether that was supposed to pervade space and time. They imagined the vortices to be knotted in various ways that explained the chemical differences between atoms of various elements. Vortices in a perfect fluid, either knotted or unknotted, would be permanent and indestructible.

Like Carter, Thomson and Tait used smoke rings as visible images of their imagined atoms. Like Carter, they failed to find any convincing evidence of a connection between image and atom. Unlike Carter, they were professional scientists, highly respected leaders of the international scientific community. Thomson was later ennobled by Queen Victoria and became Lord Kelvin, his name immortalized in the Kelvin scale of absolute temperature. Tait created a mathematical theory of knots, which grew in the twentieth century into a new branch of mathematics known as topology. Thomson and Tait were honored and respected, even as their theory of vortex atoms fell into oblivion. Wertheim asks: Why should Carter be treated differently?

In my career as a scientist, I twice had the good fortune to be a personal friend of a famous dissident. One dissident, Sir Arthur Eddington, was an insider like Thomson and Tait. The other, Immanuel Velikovsky, was an outsider like Carter. Both of them were tragic figures, intellectually brilliant and morally courageous, with the same fatal flaw as Carter. Both of them were possessed by fantasies that people with ordinary common sense could recognize as nonsense. I made it clear to both that I did not believe their fantasies, but I admired them as human beings and as imaginative artists. I admired them most of all for their stubborn refusal to remain silent. With the whole world against them, they stayed true to their beliefs. I could not pretend to agree with them, but I could give them my moral support.

Eddington was a great astronomer, one of the last of the giants who were equally gifted as observers and as theorists. His great moment as an observer came in 1919 when he led the British expedition to the island of Principe off the coast of West Africa to measure the deflection of starlight passing close to the sun during a total eclipse. The purpose of the measurement was to test Einstein's theory of general relativity. The measurement showed clearly that Einstein was right and Newton wrong. Einstein and Eddington both became immediately famous. One year later, Eddington published a book, *Space, Time and Gravitation*, that explained Einstein's ideas to English-speaking readers. It begins with a quote from Milton's *Paradise Lost*:

Perhaps to move
His laughter at their quaint opinions wide
Hereafter, when they come to model heaven
And calculate the stars: how they will wield
The mighty frame: how build, unbuild, contrive
To save appearances.

Milton had visited Galileo at his home in Florence when Galileo was under house arrest. Milton wrote poetry in Italian as well as English. He spoke Galileo's language, and used Galileo as an example in his campaign for freedom of the press in England. Milton had witnessed with Galileo the birth struggle of classical physics, as Eddington witnessed with Einstein the birth struggle of relativity three hundred years later. Eddington's book puts relativity into its proper setting as an episode in the history of Western thought. The book is marvelously clear and readable, and is probably responsible for the fact that Einstein was better understood and more admired in Britain and America than in Germany.

As a student at Cambridge University I listened to Eddington's lectures on general relativity. They were as brilliant as his books. He divided his exposition into two parts, and warned the students scrupulously when he switched from one part to the other. The first part was the orthodox mathematical theory invented by Einstein and verified by Eddington's observations. The second part was a strange concoction that he called "fundamental theory," attempting to explain all the mysteries of particle physics and cosmology with a new set of ideas. Fundamental theory was a mixture of mathematical and verbal arguments. The consequences of the theory were guessed rather than calculated. The theory had no firm basis either in physics or mathematics.

Eddington said plainly, whenever he burst into his fundamental theory with a wild rampage of speculations, "This is not generally accepted and you don't have to believe it." I was unable to decide who were more to be pitied, the bewildered students who were worried about passing the next exam or the elderly speaker who knew that he was a voice crying in the wilderness. Two facts were clear. First, Eddington was talking nonsense. Second, in spite of the nonsense, he was still a great man. For the small class of students, it was a privilege to come faithfully to his lectures and to share his pain. Two years later he was dead.

After I came to America, I became a friend of Velikovsky, who was my neighbor in Princeton. Velikovsky was a Russian Jew with an intense interest in Jewish legends and ancient history. He was born into a scholarly family in 1895 and obtained a medical degree at Moscow University in 1921. During the chaos of the Bolshevik Revolution he wrote a long Russian poem with the title "Thirty Days and Nights of Diego Pirez on the Sant'Angelo Bridge." It was published in Paris in 1935. Diego Pirez was a sixteenth-century Portuguese Jewish mystic who came to Rome and sat on the bridge near the Vatican, surrounded by beggars and thieves to whom he told his apocalyptic visions. He was condemned to death by the Inquisition, pardoned by the pope, and later burned as a heretic by the emperor Charles V.

Velikovsky escaped from Russia and settled in Palestine with his wife and daughters. He described to me the joys of practicing medicine on the slopes of Mount Carmel above Haifa, where he rode on a donkey to visit his patients in their homes. He founded and edited a journal, *Scripta Universitatis atque Bibliothecae Hierosolymitanarum*, which was the official journal of the Hebrew University before the university was established. His work for the *Scripta* was important for the founding of the Hebrew University. But he had no wish to join the university himself. To fulfill his dreams he needed complete independence. In 1939, after sixteen years in Palestine, he moved to America, where he had no license to practice medicine. To survive in America, he needed to translate his dreams into books.

Eleven years later, Macmillan published *Worlds in Collision*, and it became a best seller. Like Diego Pirez, Velikovsky told his dreams to the public in language they could understand. His dreams were mythological stories of catastrophic events, gleaned from many cultures, especially from ancient Egypt and Israel. These catastrophes were interwoven with a weird history of planetary collisions. The planets Venus and Mars were supposed to have moved out of their regular orbits and collided with the Earth a few thousand years ago.

Electromagnetic forces were invoked to counteract the normal effects of gravity. The human and cosmic events were tied together in a flowing narrative. Velikovsky wrote like an Old Testament prophet, calling down fire and brimstone from heaven, in a style familiar to Americans raised on the King James Bible. More best sellers followed: *Ages in Chaos* in 1952, *Earth in Upheaval* in 1955, *Oedipus and Akhnaton* in 1960. Velikovsky became famous as a writer and as a public speaker.

In 1977 Velikovsky asked me to write a blurb advertising his new book, *Peoples of the Sea.* I wrote a statement addressed to him personally:

> First, as a scientist, I disagree profoundly with many of the statements in your books. Second, as your friend, I disagree even more profoundly with those scientists who have tried to silence your voice. To me, you are no reincarnation of Copernicus or Galileo. You are a prophet in the tradition of William Blake, a man reviled and ridiculed by his contemporaries but now recognized as one of the greatest of English poets. A hundred and seventy years ago, Blake wrote: "The Enquiry in England is not whether a Man has Talents and Genius, but whether he is Passive and Polite and a Virtuous Ass and obedient to Noblemen's Opinions in Art and Science. If he is, he is a Good Man. If not, he must be starved." So you stand in good company. Blake, a buffoon to his enemies and an embarrassment to his friends, saw Earth and Heaven more clearly than any of them. Your poetic visions are as large as his and as deeply rooted in human experience. I am proud to be numbered among your friends.

I added the emphatic instruction, "This statement to be printed in its entirety or not at all." A quick response came from Velikovsky. He said, "How would you like it if I said you were the reincarnation of Jules Verne?" He wanted to be honored as a scientist, not as a poet. My statement was not printed, and *Peoples of the Sea* became a best seller without my help. We remained friends, and in that same year he gave me a copy of his Diego Pirez poem, which I treasure as the truest expression of his spirit. I hope it will one day be adequately translated into English.

Why do I value so highly the memory of Eddington and Velikovsky, and why does Wertheim treasure the memory of Thomson and Carter? We honor them because science is only a small part of human capability. We gain knowledge of our place in the universe not only from science but also from history, art, and literature. Science is a creative interaction of observation with imagination. "Physics at the fringe" is what happens when imagination loses touch with observation. Imagination by itself can still enlarge our vision when observation fails. The mythologies of Carter and Velikovsky fail to be science, but they are works of art and high imagining. As Blake told us long ago, "You never know what is enough unless you know what is more than enough."

Wertheim ends her book with a description of two conferences that she attended. The first was in 2003 at the Kavli Institute for Theoretical Physics at Santa Barbara. The second was in 2010 at California State University in Long Beach. Both conferences were supposed to be about physics. The subject of the Santa Barbara conference was "string cosmology." It was a gathering of the leading professional experts in the most fashionable part of theoretical physics, with David Gross, who won a Nobel Prize in 2004, presiding. Each expert in turn described a personal vision of the cosmos, delineating either a single universe or a multiplicity of universes. The various visions were incompatible with one another, and no observational evidence could prove any of them right or wrong.

The Long Beach conference was organized by the NPA, the amateurs on the fringe. Their meeting resembled a professional conference, with PowerPoint presentations followed by vigorous question-and-answer sessions. Carter was there and presented his vision of the universe among 120 others. Wertheim was probably the only person who attended both conferences. She is one of very few people who are at home in both worlds. She is a professional science writer with a degree in physics, and she has made friends with many insiders as well as outsiders. She asks at the end the central question raised by her book: Why should we pay more attention to one set of self-proclaimed experts than to the other?

So far as science in general is concerned, the answer to Wertheim's question is clear. There is good reason to pay more attention to scientific experts than to amateurs, so long as science is based on experiments. Only trained experts can do experiments with the care and precision that experiments demand. Expert experimenters are not infallible, but they are less fallible than amateurs. Experiments give orthodox beliefs a solid basis. An experimental basis exists for the established disciplines of physics and chemistry and biology. However, some parts of physics are less secure than others, because the experts in physics are divided into experimenters and theorists.

Over most of the territory of physics, theorists and experimenters are engaged in a common enterprise, and theories are tested rigorously by experiment. The theorists listen to the voice of nature speaking through experimental tools. This was true for the great theorists of the early twentieth century, Einstein and Heisenberg and Schrödinger, whose revolutionary theories of relativity and quantum mechanics were tested by precise experiments and found to fit the facts of nature. The new mathematical abstractions fit the facts, while the old mechanical models did not.

String cosmology is different. String cosmology is a part of theoretical physics that has become detached from experiments. String cosmologists are free to imagine universes and multiverses, guided by intuition and aesthetic judgment alone. Their creations must be logically consistent and mathematically elegant, but they are otherwise unconstrained. That is why Wertheim found the official string cosmology conference disconcertingly similar to the unofficial NPA conference. The insiders and the outsiders seem to be following the same rules. Both groups are telling stories of imagined worlds, and neither has an assured way of deciding who is right. If the title *Physics on the Fringe* fits the natural philosophers, the same title also fits the string cosmologists.

The fringe of physics is not a sharp boundary with truth on one side and fantasy on the other. All of science is uncertain and subject to revision. The glory of science is to imagine more than we can prove. The fringe is the unexplored territory where truth and fantasy are not yet disentangled. Hermann Weyl, who was one of the main architects of the relativity and quantum revolutions, said to me once, "I always try to combine the true with the beautiful, but when I have to choose one or the other, I usually choose the beautiful." Following Weyl's good example, our string cosmologists are making the same choice.

Note added in 2014: Velikovsky's elder daughter, Shulamit Kogan, settled with her husband in Israel; the younger daughter, Ruth Sharon, settled in Princeton. Both daughters were fiercely loyal to their father. Ruth took care of the Velikovsky archive until 2005, when she presented it to the Princeton University Library. She published a biography, Aba, the Glory and the Torment: The Life of Dr. Immanuel Velikovsky *(Paradigma, 2010).*

14

HOW WE KNOW

THE FIRST CHAPTER of James Gleick's *The Information* has the title "Drums That Talk."* It explains the concept of information by looking at a simple example. The example is a drum language used in a part of the Democratic Republic of Congo where the human language is Kele. European explorers had been aware for a long time that the irregular rhythms of African drums were carrying mysterious messages through the jungle. Explorers would arrive at villages where no European had been before and find that the village elders were already prepared to meet them.

Sadly, the drum language was only understood and recorded by a single European before it started to disappear. The European was John Carrington, an English missionary who spent his life in Africa and became fluent in both Kele and drum language. He arrived in Africa in 1938 and published his findings in 1949 in a book, *The Talking Drums of Africa*.† Before the arrival of the Europeans with their roads and radios, the Kele-speaking Africans had used the drum language for rapid communication from village to village in

---

* *The Information: A History, a Theory, a Flood* (Pantheon, 2011).

†London: Carey Ringsgate, 1949.

the rain forest. Every village had an expert drummer and every villager could understand what the drums were saying. By the time Carrington wrote his book, the use of drum language was already fading and schoolchildren were no longer learning it. In the sixty years since then, telephones made drum language obsolete and completed the process of extinction.

Carrington understood how the structure of the Kele language made drum language possible. Kele is a tonal language with two sharply distinct tones. Each syllable is either low or high. The drum language is spoken by a pair of drums with the same two tones. Each Kele word is spoken by the drums as a sequence of low and high beats. In passing from human Kele to drum language, all the information contained in vowels and consonants is lost. In a European language, the consonants and vowels contain all the information, and if this information were dropped there would be nothing left. But in a tonal language like Kele, some information is carried in the tones and survives the transition from human speaker to drums. The fraction of information that survives in a drum word is small, and the words spoken by the drums are correspondingly ambiguous. A single sequence of tones may have hundreds of meanings depending on the missing vowels and consonants. The drum language must resolve the ambiguity of the individual words by adding more words. When enough redundant words are added, the meaning of the message becomes unique.

In 1954 a visitor from the United States came to Carrington's mission school. Carrington was taking a walk in the forest and his wife wished to call him home for lunch. She sent him a message in drum language and explained it to the visitor. To be intelligible to Carrington, the message needed to be expressed with redundant and repeated phrases: "White man spirit in forest come come to house of shingles high up above of white man spirit in forest. Woman with yam awaits. Come come." Carrington heard the message and came home. On the average, about eight words of drum language were needed to transmit one word of human language unambiguously. Western mathematicians would say that about one eighth of the information in the human Kele language belongs to the tones that are transmitted by the drum language. The redundancy of the drum language phrases compensates for the loss of the information in vowels and consonants. The African drummers knew nothing of Western mathematics, but they found the right level of redundancy for their drum language by trial and error. Carrington's wife had learned the language from the drummers and knew how to use it.

The story of the drum language illustrates the central dogma of information theory. The central dogma says, "Meaning is irrelevant." Information is independent of the meaning that it expresses and of the language used to express it. Information is an abstract concept, which can be embodied equally well in human speech or in writing or in drumbeats. All that is needed to transfer information from one language to another is a coding system. A coding system may be simple or complicated. If the code is simple, as it is for the drum language with its two tones, a given amount of information requires a longer message. If the code is complicated, as it is for spoken language, the same amount of information can be conveyed in a shorter message.

Another example illustrating the central dogma is the French optical telegraph. Until 1793, the fifth year of the French Revolution, the African drummers were ahead of Europeans in their ability to transmit information rapidly over long distances. In 1793, Claude Chappe, a patriotic citizen of France, wishing to strengthen the defense of the revolutionary government against domestic and foreign enemies, invented a device that he called the telegraph. The telegraph was an optical communication system with stations consisting of large movable pointers mounted on the tops of sixty-foot towers. Each station was manned by an operator who could read a message transmitted by a neighboring station and transmit the same message to the next station in the transmission line.

The distance between neighbors was about seven miles. Along the transmission lines, optical messages in France could travel faster than drum messages in Africa. When Napoleon took charge of the French Republic in 1799, he ordered the completion of the optical telegraph system to link all the major cities of France from Calais and Paris to Toulon and onward to Milan. The telegraph became, as Chappe had intended, an important instrument of national power. Napoleon made sure that it was not available to private users.

Unlike the drum language, which was based on spoken language, the optical telegraph was based on written French. Chappe invented an elaborate coding system to translate written messages into optical signals. He had the opposite problem from the drummers. The drummers had a fast transmission system with ambiguous messages. They needed to slow down the transmission to make the messages unambiguous. Chappe had a painfully slow transmission system with redundant messages. The French language, like most alphabetic languages, is highly redundant, using many more letters than are needed to convey the meaning of a message. Chappe's coding system allowed messages to be transmitted faster. Many common phrases and proper names were encoded by only two optical symbols, with a substantial gain in speed of transmission. The composer and the reader of the message had codebooks listing the message codes for eight thousand phrases and names. For Napoleon it was an advantage to have a code that was effectively cryptographic, keeping the content of the messages secret from citizens along the route.

After these two historical examples of rapid communication in Africa and France, the rest of Gleick's book is about the modern development of information technology. The modern history is dominated by two Americans, Samuel Morse and Claude Shannon. Morse was the inventor of Morse code. He was also one of the pioneers who built a telegraph system using electricity conducted through wires instead of optical pointers deployed on towers. Morse launched his electric telegraph in 1838 and perfected the code in 1844. His code used short and long pulses of electric current to represent letters of the alphabet.

Morse was ideologically at the opposite pole from Chappe. He was not interested in secrecy or in creating an instrument of government power. The Morse system was designed to be a profit-making enterprise, fast and cheap and available to everybody. At the beginning the price of a message was a quarter of a cent per letter. The most important users of the system were newspaper correspondents spreading news of local events to readers all over the world. Morse code was simple enough that anyone could learn it. The system provided no secrecy to the users. If users wanted secrecy, they could invent their own secret codes and encipher their messages themselves. The price of a message in cipher was higher than the price of a message in plain text, because the telegraph operators could transcribe plain text faster. It was much easier to correct errors in plain text than in cipher.

Shannon was the founding father of information theory. For a hundred years after the electric telegraph, other communication systems such as the telephone, radio, and television were invented and developed by engineers without any need for higher mathematics. Then Shannon supplied the theory to understand all of these systems together, defining information as an abstract quantity inherent in a telephone message or a television picture. He brought higher mathematics into the game.

When Shannon was a boy growing up on a farm in Michigan, he

built a homemade telegraph system using Morse code. Messages were transmitted to friends on neighboring farms, using the barbed wire of their fences to conduct electric signals. When World War II began, Shannon became one of the pioneers of scientific cryptography, working on the high-level cryptographic telephone system that allowed Roosevelt and Churchill to talk to each other over a secure channel. Shannon's friend Alan Turing was also working as a cryptographer at the same time, in the famous British Enigma project that successfully deciphered German military codes. The two pioneers met frequently when Turing visited New York in 1943, but they belonged to separate secret worlds and could not exchange ideas about cryptography.

In 1945 Shannon wrote a paper, "A Mathematical Theory of Cryptography," which was stamped SECRET and never saw the light of day. He published in 1948 an expurgated version of the 1945 paper with the title "A Mathematical Theory of Communication." The 1948 version appeared in the *Bell System Technical Journal*, the house journal of Bell Telephone Laboratories, and became an instant classic. It is the founding document for the modern science of information. After Shannon, the technology of information raced ahead, with electronic computers, digital cameras, the Internet, and the World Wide Web.

According to Gleick, the impact of information on human affairs came in three installments: first, the history, the thousands of years during which people created and exchanged information without the concept of measuring it; second, the theory, first formulated by Shannon; third, the flood, in which we now live. The flood began quietly. The event that made the flood plainly visible occurred in 1965, when Gordon Moore stated Moore's law. Moore was an electrical engineer, the founder of Intel Corporation, a company that manufactured components for computers and other electronic gadgets. His law said that the price of electronic components would decrease and their numbers would increase by a factor of two every eighteen months. This implied that the price would decrease and the numbers would increase by a factor of a hundred every decade. Moore's prediction of continued growth has turned out to be astonishingly accurate during the forty-five years since he announced it. In these four and a half decades, the price has decreased and the numbers have increased by a factor of a billion, nine powers of ten. Nine powers of ten are enough to turn a trickle into a flood.

Moore was in the hardware business, making hardware components for electronic machines, and he stated his law as a law of growth for hardware. But the law applies also to the information that the hardware is designed to embody. The purpose of the hardware is to store and process information. The storage of information is called memory, and the processing of information is called computing. The consequence of Moore's law for information is that the price of memory and computing decreases and the available amount of memory and computing increases by a factor of a hundred every decade. The flood of hardware becomes a flood of information.

In 1949, one year after Shannon published the rules of information theory, he drew up a table of the various stores of memory that then existed. The biggest memory in his table was the Library of Congress, which he estimated to contain one hundred trillion bits of information. That was at the time a fair guess at the sum total of recorded human knowledge. Today a memory disk drive storing that amount of information weighs a few pounds and can be bought for about a thousand dollars. Information, otherwise known as data, pours into memories of that size or larger, in government and business offices and scientific laboratories all over the world. Gleick quotes the computer scientist Jaron Lanier describing the effect of the flood: "It's as if you kneel to plant the seed of a tree and it grows so fast that it swallows your whole town before you can even rise to your feet."

On December 8, 2010, Gleick published on the *The New York Review*'s blog an illuminating essay, "The Information Palace." It was written too late to be included in his book. It describes the historical changes of meaning of the word "information," as recorded in the latest quarterly online revision of the *Oxford English Dictionary.* The word first appears in 1386 in a parliamentary report with the meaning "denunciation." The history ends with the modern usage, "information fatigue," defined as "apathy, indifference or mental exhaustion arising from exposure to too much information."

The consequences of the information flood are not all bad. One of the creative enterprises made possible by the flood is Wikipedia, started ten years ago by Jimmy Wales. Among my friends and acquaintances, everybody distrusts Wikipedia and everybody uses it. Distrust and productive use are not incompatible. Wikipedia is the ultimate open-source repository of information. Everyone is free to read it and everyone is free to write it. It contains articles in 262 languages written by several million authors. The information that it contains is totally unreliable and surprisingly accurate. It is often unreliable because many of the authors are ignorant or careless. It is often accurate because the articles are edited and corrected by readers who are better informed than the authors.

Wales hoped when he started Wikipedia that the combination of enthusiastic volunteer writers with open-source information technology would cause a revolution in human access to knowledge. The rate of growth of Wikipedia exceeded his wildest dreams. Within ten years it has become the biggest storehouse of information on the planet and the noisiest battleground of conflicting opinions. It illustrates Shannon's law of reliable communication. Shannon's law says that accurate transmission of information is possible in a communication system with a high level of noise. Even in the noisiest system, errors can be reliably corrected and accurate information transmitted, provided that the transmission is sufficiently redundant. That is, in a nutshell, how Wikipedia works.

The information flood has also brought enormous benefits to science. The public has a distorted view of science because children are taught in school that science is a collection of firmly established truths. In fact, science is not a collection of truths. It is a continuing exploration of mysteries. Wherever we go exploring in the world around us, we find mysteries. Our planet is covered by continents and oceans whose origin we cannot explain. Our atmosphere is constantly stirred by poorly understood disturbances that we call weather and climate. The visible matter in the universe is outweighed by a much larger quantity of dark invisible matter that we do not understand at all. The origin of life is a total mystery, and so is the existence of human consciousness. We have no clear idea how the electrical discharges occurring in nerve cells in our brains are connected with our feelings and desires and actions.

Even physics, the most exact and most firmly established branch of science, is still full of mysteries. We do not know how much of Shannon's theory of information will remain valid when quantum devices replace classical electric circuits as the carriers of information. Quantum devices may be made of single atoms or microscopic magnetic circuits. All that we know for sure is that they can theoretically do certain jobs that are beyond the reach of classical devices. Quantum computing is still an unexplored mystery on the frontier of information theory. Science is the sum total of a great multitude of mysteries. It is an unending argument between a great multitude of voices. Science resembles Wikipedia much more than it resembles the *Encyclopaedia Britannica*.

The rapid growth of the flood of information in the last ten years made Wikipedia possible, and the same flood made twenty-first-century science possible. Twenty-first-century science is dominated by huge stores of information that we call databases. The information flood has made it easy and cheap to build databases. One example of a twenty-first-century database is the collection of genome sequences of living creatures belonging to various species from microbes to humans. Each genome contains the complete genetic information that shaped the creature to which it belongs. The genome database is rapidly growing and is available for scientists all over the world to explore. Its origin can be traced to 1939, when Shannon wrote his PhD thesis, "An Algebra for Theoretical Genetics."

Shannon was then a graduate student in the mathematics department at MIT. He was only dimly aware of the possible physical embodiment of genetic information. The true physical embodiment of the genome is the double-helix structure of DNA molecules, discovered by Francis Crick and James Watson fourteen years later. In 1939 Shannon understood that the basis of genetics must be information, and that the information must be coded in some abstract algebra independent of its physical embodiment. Without any knowledge of the double helix, he could not hope to guess the detailed structure of the genetic code. He could only imagine that in some distant future the genetic information would be decoded and collected in a giant database that would define the total diversity of living creatures. It took only sixty years for his dream to come true.

In the twentieth century, genomes of humans and other species were laboriously decoded and translated into sequences of letters in computer memories. The decoding and translation became cheaper and faster as time went on, the price decreasing and the speed increasing according to Moore's law. The first human genome took fifteen years to decode and cost about a billion dollars. Now a human genome can be decoded in a few weeks and costs a few thousand dollars. Around the year 2000, a turning point was reached, when it became cheaper to produce genetic information than to understand it. Now we can pass a piece of human DNA through a machine and rapidly read out the genetic information, but we cannot read out the meaning of the information. We shall not fully understand the information until we understand in detail the processes of embryonic development that the DNA orchestrated to make us what we are.

A similar turning point was reached about the same time in the science of astronomy. Telescopes and spacecraft have evolved slowly, but cameras and optical data processors have evolved fast. Modern sky-survey projects collect data from huge areas of sky and produce databases with accurate information about billions of objects. Astronomers without access to large instruments can make discoveries by mining the databases instead of observing the sky. Big databases have caused similar revolutions in other sciences such as biochemistry and ecology.

The explosive growth of information in our human society is a part of the slower growth of ordered structures in the evolution of life as a whole. Life has for billions of years been evolving with organisms and ecosystems embodying increasing amounts of information. The evolution of life is a part of the evolution of the universe, which also evolves with increasing amounts of information embodied in ordered structures: galaxies and stars and planetary systems. In the living and in the nonliving world, we see a growth of order, starting from the featureless and uniform gas of the early universe and producing the magnificent diversity of weird objects that we see in the sky and in the rain forest. Everywhere around us, wherever we look, we see evidence of increasing order and increasing information. The technology arising from Shannon's discoveries is only a local acceleration of the natural growth of information.

The visible growth of ordered structures in the universe seemed paradoxical to nineteenth-century scientists and philosophers, who believed in a dismal doctrine called the heat death. Lord Kelvin, one of the leading physicists of that time, promoted the heat death dogma, predicting that the flow of heat from warmer to cooler objects will result in a decrease of temperature differences everywhere, until all temperatures ultimately become equal. Life needs temperature differences to avoid being stifled by its waste heat. So life will disappear.

This dismal view of the future was in startling contrast to the ebullient growth of life that we see around us. Thanks to the discoveries of astronomers in the twentieth century, we now know that the heat death is a myth. The heat death can never happen, and there is no paradox. The best popular account of the disappearance of the paradox is a chapter, "How Order Was Born of Chaos," in the book *Creation of the Universe*, by Fang Lizhi and his wife, Li Shuxian.* Fang is doubly famous as a leading Chinese astronomer and a leading political dissident. He is now pursuing his double career at the University of Arizona.

The belief in a heat death was based on an idea that I call the cooking rule. The cooking rule says that a piece of steak gets warmer when we put it on a hot grill. More generally, the rule says that any object gets warmer when it gains energy and gets cooler when it loses energy. Humans have been cooking steaks for thousands of years, and nobody ever saw a steak get colder while cooking on a fire. The cooking rule is true for objects small enough for us to handle. If the cooking rule is always true, then Lord Kelvin's argument for the heat death is correct.

We now know that the cooking rule is not true for objects of astronomical size, for which gravitation is the dominant form of en-

---

*Singapore: World Scientific Publishing Co., 1989.

ergy. The sun is a familiar example. As the sun loses energy by radiation, it becomes hotter and not cooler. Since the sun is made of compressible gas squeezed by its own gravitation, loss of energy causes it to become smaller and denser, and the compression causes it to become hotter. For almost all astronomical objects, gravitation dominates, and they have the same unexpected behavior. Gravitation reverses the usual relation between energy and temperature. In the domain of astronomy, when heat flows from hotter to cooler objects, the hot objects get hotter and the cool objects get cooler. As a result, temperature differences in the astronomical universe tend to increase rather than decrease as time goes on. There is no final state of uniform temperature, and there is no heat death. Gravitation gives us a universe hospitable to life. Information and order can continue to grow for billions of years in the future, as they have evidently grown in the past.

The vision of the future as an infinite playground, with an unending sequence of mysteries to be understood by an unending sequence of players exploring an unending supply of information, is a glorious vision for scientists. Scientists find the vision attractive, since it gives them a purpose for their existence and an unending supply of jobs. The vision is less attractive to artists and writers and ordinary people. Ordinary people are more interested in friends and family than in science. Ordinary people may not welcome a future spent swimming in an unending flood of information. A darker view of the information-dominated universe was described in the famous story "The Library of Babel," written by Jorge Luis Borges in 1941.* Borges imagined his library, with an infinite array of books and shelves and mirrors, as a metaphor for the universe.

Gleick's book has an epilogue entitled "The Return of Meaning,"

**Labyrinths: Selected Stories and Other Writings* (New Directions, 1962).

expressing the concerns of people who feel alienated from the prevailing scientific culture. The enormous success of information theory came from Shannon's decision to separate information from meaning. His central dogma, "Meaning is irrelevant," declared that information could be handled with greater freedom if it was treated as a mathematical abstraction independent of meaning. The consequence of this freedom is the flood of information in which we are drowning. The immense size of modern databases gives us a feeling of meaninglessness. Information in such quantities reminds us of Borges's library extending infinitely in all directions. It is our task as humans to bring meaning back into this wasteland. As finite creatures who think and feel, we can create islands of meaning in the sea of information. Gleick ends his book with Borges's image of the human condition:

> We walk the corridors, searching the shelves and rearranging them, looking for lines of meaning amid leagues of cacophony and incoherence, reading the history of the past and of the future, collecting our thoughts and collecting the thoughts of others, and every so often glimpsing mirrors, in which we may recognize creatures of the information.

Note added in 2014: Fang Lizhi died in 2012 at the age of seventy-six. Until the end he remained active in his double life as astronomical thinker and political dissident.

Two corrections to the review: First, the British Enigma project, which deciphered German military codes in World War II, started with crucial help from Polish cryptologists. Before the war began in 1939, the Poles captured a German Enigma machine and gave copies of it to Britain and France. To have the machine was an essential first step toward deciphering the codes. Second, Borges's "The Library of Babel" was not infinite. The number of books was finite but too large to be counted. I thank two vigilant readers for these corrections.

15

THE "DRAMATIC PICTURE" OF RICHARD FEYNMAN

IN THE LAST hundred years, since radio and television created the modern worldwide mass-market entertainment industry, there have been two scientific superstars, Albert Einstein and Stephen Hawking. Lesser lights such as Carl Sagan and Neil deGrasse Tyson and Richard Dawkins have a big public following, but they are not in the same class as Einstein and Hawking. Sagan, Tyson, and Dawkins have fans who understand their message and are excited by their science. Einstein and Hawking have fans who understand almost nothing about science and are excited by their personalities.

On the whole, the public shows good taste in its choice of idols. Einstein and Hawking earned their status as superstars not only by their scientific discoveries but by their outstanding human qualities. Both of them fit easily into the role of icon, responding to public adoration with modesty and good humor and with provocative statements calculated to command attention. Both of them devoted their lives to an uncompromising struggle to penetrate the deepest mysteries of nature, and both still had time left over to care about the practical worries of ordinary people. The public rightly judged them to be genuine heroes, friends of humanity as well as scientific wizards.

Two new books now raise the question whether Richard Feynman is rising to the status of superstar. The books are very different in style and in substance. Lawrence Krauss's *Quantum Man* is a narrative of Feynman's life as a scientist, skipping lightly over the personal adventures that have been emphasized in earlier biographies.* Krauss succeeds in explaining in nontechnical language the essential core of Feynman's thinking. Unlike any previous biographer, he takes the reader inside Feynman's head and reconstructs the picture of nature as Feynman saw it. This is a new kind of scientific history, and Krauss is well qualified to write it, being an expert physicist and a gifted writer of scientific books for the general public. *Quantum Man* shows us the side of Feynman's personality that was least visible to most of his admirers, the silent and persistent calculator working intensely through long days and nights to figure out how nature behaves.

The other book, *Feynman* by writer Jim Ottaviani and artist Leland Myrick, is very different.† It is a comic-book biography of Feynman, containing 266 pages of pictures of Feynman and his legendary adventures. In every picture, bubbles of text record Feynman's comments, mostly taken from stories that he and others had told and published in earlier books. We see Feynman first as an inquisitive five-year-old, learning from his father to question authority and admit ignorance. He asks his father at the playground, "Why does [the ball] keep moving?" His father says, "The reason the ball keeps rolling is because it has 'inertia.' That's what scientists call the reason . . . , but it's just a name. Nobody really knows what it means." His father was a traveling salesman without scientific training, but he understood the difference between giving a thing a name and knowing

**Quantum Man: Richard Feynman's Life in Science* (Norton, 2011).

†First Second, 2011.

how it works. He ignited in his son a lifelong passion to know how things work.

After the scenes with his father, the pictures show Feynman changing gradually through the roles of ebullient young professor and carnival drum-player, doting parent and loving husband, revered teacher and educational reformer, until he ends his life as a wrinkled sage in a losing battle with cancer. It comes as a shock to see myself portrayed in these pages, as a lucky young student taking a four-day ride with Feynman in his car from Cleveland to Albuquerque, sharing with him some unusual lodgings and entertained by an unending stream of his memorable conversation.

One of the incidents in Feynman's life that displayed his human qualities sharply was his reaction to the news in 1965 that he had won a Nobel Prize. When the telephone call came from Stockholm, he made remarks that appeared arrogant and ungrateful. He said he would probably refuse the prize, since he hated formal ceremonies and particularly hated the pompous rituals associated with kings and queens. His father had told him when he was a kid, "What are kings anyway? Just guys in fancy clothes." He would rather refuse the prize than be forced to dress up and shake hands with the king of Sweden.

But after a few days, he changed his mind and accepted the prize. As soon as he arrived in Sweden, he made friends with the Swedish students who came to welcome him. At the banquet when he officially accepted the prize, he gave an impromptu speech, apologizing for his earlier rudeness and thanking the Swedish people with a moving personal account of the blessings that the prize had brought to him.

Feynman had looked forward to meeting Sin-Itiro Tomonaga, the Japanese physicist who shared the Nobel Prize with him. Tomonaga had independently made some of the same discoveries as Feynman, five years earlier, in the total isolation of wartime Japan. He shared with Feynman not only ideas about physics but also experiences of personal tragedy. In the spring of 1945, Feynman was nursing his beloved first wife, Arline, through the last weeks of her life as he watched her die from tuberculosis. In the same spring, Tomonaga was helping a group of his students to survive in the ashes of Tokyo, after a firestorm devastated the city and killed an even greater number of people than the nuclear bomb would kill in Hiroshima four months later. Feynman and Tomonaga shared three outstanding qualities: emotional toughness, intellectual integrity, and a robust sense of humor.

To Feynman's dismay, Tomonaga failed to appear in Stockholm. The Ottaviani-Myrick book has Tomonaga explaining what happened:

> Although I sent a letter saying that I would be "pleased to attend," I loathed the thought of going, thinking that the cold would be severe, as the ceremony was to be held in December, and that the inevitable formalities would be tiresome. After I was named a Nobel Prize awardee, many people came to visit, bringing liquor. I had barrels of it. One day, my father's younger brother, who loved whiskey, happened to stop by and we both began drinking gleefully. We drank a little too much, and then, seizing the opportunity that my wife had gone out shopping, I entered the bathroom to take a bath. There I slipped and fell down, breaking six of my ribs.... It was a piece of good luck in that unhappy incident.

After Tomonaga recovered from his injuries, he was invited to England to receive another high honor requiring a formal meeting with royalty. This time he did not slip in the bathtub. He duly appeared at Buckingham Palace to shake hands with the English queen. The queen did not know that he had failed to travel to Stockholm. She innocently asked him whether he had enjoyed his meeting with the king of Sweden. Tomonaga was totally flummoxed. He could not bring himself to confess to the queen that he had gotten drunk and broken his ribs. He said that he had enjoyed his conversation with the king very much. He remarked afterward that for the rest of his life he would be carrying a double burden of guilt: first for getting drunk, and second for telling a lie to the queen of England.

Twenty years later, when Feynman was mortally ill with cancer, he served on the NASA commission investigating the *Challenger* disaster of 1986. He undertook this job reluctantly, knowing that it would use up most of the time and strength that he had left. He undertook it because he felt an obligation to find the root causes of the disaster and to speak plainly to the public about his findings. He went to Washington and found what he had expected at the heart of the tragedy: a bureaucratic hierarchy with two groups of people, the engineers and the managers, who lived in separate worlds and did not communicate with each other. The engineers lived in the world of technical facts; the managers lived in the world of political dogmas.

He asked members of both groups to tell him their estimates of the risk of disastrous failure in each Space Shuttle mission. The engineers estimated the risk to be of the order of one disaster in a hundred missions. The managers estimated the risk to be of the order of one disaster in a hundred thousand missions. The difference, a factor of a thousand between the two estimates, was never reconciled and never openly discussed. The managers were in charge of the operations and made the decisions to fly or not to fly, based on their own estimates of the risk. But the technical facts that Feynman uncovered proved that the managers were wrong and the engineers were right.

Feynman had two opportunities to educate the public about the causes of the disaster. The first opportunity concerned the technical facts. An open meeting of the commission was held with newspaper and television reporters present. Feynman was prepared with a glass of ice water and a sample of a rubber O-ring seal from a shuttle solid-fuel booster rocket. He dipped the piece of rubber into the ice water, pulled it out, and demonstrated the fact that the cold rubber was stiff. The cold rubber would not function as a gas-tight seal to keep the hot rocket exhaust away from the structure. Since the *Challenger* launch had occurred on January 28 in unusually cold weather, Feynman's little demonstration pointed to the stiffening of the O-ring seal as a probable technical cause of the disaster.

The second opportunity to educate the public concerned the culture of NASA. Feynman wrote an account of the cultural situation as he saw it, with the fatal division of the NASA administration into two noncommunicating cultures: engineers and managers. The political dogma of the managers, declaring risks to be a thousand times smaller than the technical facts would indicate, was the cultural cause of the disaster. The political dogma arose from a long history of public statements by political leaders that the shuttle was safe and reliable. Feynman ended his account with the famous declaration: "For a successful technology, reality must take precedence over public relations, for nature cannot be fooled."

Feynman fought hard to have his statement of conclusions incorporated in the official report of the commission. The chairman of the commission, William Rogers, was a professional politician with long experience in government. Rogers wished the public to believe that the *Challenger* disaster was a highly unlikely accident for which NASA was not to blame. He fought hard to exclude Feynman's statement from the report. In the end a compromise was reached. Feynman's statement was not included in the report but was added as an appendix at the end, with a note saying that it was Feynman's personal statement and not agreed to by the commission. This compromise worked to Feynman's advantage. As he remarked at the time, the appendix standing at the end got much more public attention than it would have if it had been part of the official report.

Feynman's dramatic exposure of NASA incompetence and his O-ring demonstrations made him a hero to the general public. The event was the beginning of his rise to the status of superstar. Before his service on the *Challenger* commission, he was widely admired by knowledgeable people as a scientist and a colorful character. Afterward, he was admired by a much wider public, as a crusader for honesty and plain speaking in government. Anyone fighting secrecy and corruption in any part of the government could look to Feynman as a leader.

In the final scene of the comic book, Feynman is walking on a mountain trail with his friend Danny Hillis. Hillis says, "I'm sad because you're going to die." Feynman replies, "Yeah, that bugs me sometimes too. But not as much as you think. See, when you get as old as I am, you start to realize that you've told most of the good stuff you know to other people anyway. Hey! I bet I can show you a better way home." And Hillis is left alone on the mountain. These images capture with remarkable sensitivity the essence of Feynman's character. The comic-book picture somehow comes to life and speaks with the voice of the real Feynman.

Twenty years ago, when I was traveling on commuter trains in the suburbs of Tokyo, I was astonished to see that a large fraction of the Japanese commuters were reading books, and that a large fraction of the books were comic books. The genre of serious comic-book literature was highly developed in Japan long before it appeared in the West. The Ottaviani-Myrick book is the best example of this genre that I have yet seen with text in English. Some Western readers commonly use the Japanese word *manga* to mean serious comic-book literature. According to one of my Japanese friends, this usage is wrong. The word *manga* means "idle picture" and is used in Japan to describe collections of trivial comic-book stories. The correct word for serious comic-book literature is *gekiga*, meaning "dramatic picture." The Feynman picture book is a fine example of *gekiga* for Western readers.

The title of Krauss's book, *Quantum Man*, is well chosen. The central theme of Feynman's work as a scientist was to explore a new way of thinking and working with quantum mechanics. The book succeeds in explaining without any mathematical jargon how Feynman thought and worked. This is possible because he visualized the world with pictures rather than with equations. Other physicists in the past and present describe the laws of nature with equations and then solve the equations to find out what happens. Feynman skipped the equations and wrote down the solutions directly, using his pictures as a guide. Skipping the equations was his greatest contribution to science. By skipping the equations, he created the language that a majority of modern physicists speak. Incidentally, he created a language that ordinary people without mathematical training can understand. To use the language to do quantitative calculations requires training, but untrained people can use it to describe qualitatively how nature behaves.

Feynman's picture of the world starts from the idea that the world has two layers: a classical layer and a quantum layer. Classical means that things are ordinary. Quantum means that things are weird. We live in the classical layer. All the things that we can see and touch and measure, such as bricks and people and energies, are classical. We see them with classical devices such as eyes and cameras, and we measure them with classical instruments such as thermometers and clocks. The pictures that Feynman invented to describe the world are classical pictures of objects moving in the classical layer. Each picture represents a possible history of the classical layer. But the real world of atoms and particles is not classical. Atoms and particles appear in Feynman's pictures as classical objects, but they actually obey quite different laws. They obey the quantum laws that Feynman showed us how to describe by using his pictures. The world of atoms belongs to the quantum layer, which we cannot touch directly.

The primary difference between the classical layer and the quantum layer is that the classical layer deals with facts and the quantum layer deals with probabilities. In situations where classical laws are valid, we can predict the future by observing the past. In situations where quantum laws are valid, we can observe the past but we cannot predict the future. In the quantum layer, events are unpredictable. The Feynman pictures only allow us to calculate the probabilities that various alternative futures may happen.

The quantum layer is related to the classical layer in two ways. First, the state of the quantum layer is what is called "a sum over histories," that is, a combination of every possible history of the classical layer leading up to that state. Each possible classical history is given a quantum amplitude. The quantum amplitude, otherwise known as a wave function, is a number defining the contribution of that classical history to that quantum state. Second, the quantum amplitude is obtained from the picture of that classical history by following a simple set of rules. The rules are pictorial, translating the picture directly into a number. The difficult part of the calculation is to add up the sum over histories correctly. The great achievement of Feynman was to show that this sum-over-histories view of the quantum world reproduces all the known results of quantum theory and allows an exact description of quantum processes in situations where earlier versions of quantum theory had broken down.

Feynman was radical in his disrespect for authority but conservative in his science. When he was young he had hoped to start a revolution in science, but nature said no. Nature told him that the existing jungle of scientific ideas, with the classical world and the quantum world described by very different laws, was basically correct. He tried to find new laws of nature, but the result of his efforts was in the end to consolidate the existing laws in a new structure. He hoped to find discrepancies that would prove the old theories wrong, but nature stubbornly persisted in proving them right. However disrespectful he might be to famous old scientists, he was never disrespectful to nature.

Toward the end of Feynman's life, his conservative view of quantum science became unfashionable. The fashionable theorists reject his dualistic picture of nature, with the classical world and the quantum world existing side by side. They believe that only the quantum world is real, and the classical world must be explained as some kind of illusion arising out of quantum processes. They disagree about the way in which quantum laws should be interpreted. Their basic problem is to explain how a world of quantum probabilities can generate the illusions of classical certainty that we experience in our daily lives. Their various interpretations of quantum theory lead to competing philosophical speculations about the role of the observer in the description of nature.

Feynman had no patience for such speculations. He said that nature tells us that both the quantum world and the classical world exist and are real. We do not understand precisely how they fit together. According to Feynman, the road to understanding is not to argue about philosophy but to continue exploring the facts of nature. In recent years, a new generation of experimenters has been advancing along Feynman's road with great success, moving into the new worlds of quantum computing and quantum cryptography.

Krauss shows us a portrait of a scientist who was unusually unselfish. His disdain for honors and rewards was genuine. After he was elected to membership of the National Academy of Sciences, he resigned his membership because the members of the academy spent too much of their time debating who was worthy of admission in the next academy election. He considered the academy to be more concerned with self-glorification than with public service. He hated all hierarchies, and wanted no badge of superior academic status to come between him and his younger friends. He considered science to be a collective enterprise in which educating the young was as important as making personal discoveries. He put as much effort into his teaching as into his thinking.

He never showed the slightest resentment when I published some of his ideas before he did. He told me that he avoided disputes about priority in science by following a simple rule: "Always give the bastards more credit than they deserve." I have followed this rule myself. I find it remarkably effective for avoiding quarrels and making friends. A generous sharing of credit is the quickest way to build a healthy scientific community. In the end, Feynman's greatest contribution to science was not any particular discovery. His contribution was the creation of a new way of thinking that enabled a great multitude of students and colleagues, including me, to make their own discoveries.

16

HOW TO DISPEL YOUR ILLUSIONS

IN 1955, WHEN Daniel Kahneman was twenty-one years old, he was a lieutenant in the Israel Defense Forces. He was given the job of setting up a new interview system for the entire army. The purpose was to evaluate each freshly drafted recruit and put him or her into the appropriate slot in the war machine. The interviewers were supposed to predict who would do well in the infantry or the artillery or the tank corps or the various other branches of the army. The old interview system, before Kahneman arrived, was informal. The interviewers chatted with the recruit for fifteen minutes and then came to a decision based on the conversation. The system had failed miserably. When the actual performance of the recruit a few months later was compared with the performance predicted by the interviewers, the correlation between actual and predicted performance was zero.

Kahneman had a bachelor's degree in psychology and had read *Clinical vs. Statistical Prediction: A Theoretical Analysis and a Review of the Evidence* by Paul Meehl, published only a year earlier. Meehl was an American psychologist who studied the successes and failures of predictions in many different settings. He found overwhelming evidence for a disturbing conclusion. Predictions based on simple statistical scoring were generally more accurate than predictions based on expert judgment.

A famous example confirming Meehl's conclusion is the Apgar score, invented by the anesthesiologist Virginia Apgar in 1953 to guide the treatment of newborn babies. The Apgar score is a simple formula based on five vital signs that can be measured quickly: heart rate, breathing, reflexes, muscle tone, and color. It does better than the average doctor in deciding whether the baby needs immediate help. It is now used everywhere and saves the lives of thousands of babies. Another famous example of statistical prediction is the Dawes formula for the durability of marriage. The formula is "frequency of love-making minus frequency of quarrels." Robyn Dawes was a psychologist who worked with Kahneman later. His formula does better than the average marriage counselor in predicting whether a marriage will last.

Having read the Meehl book, Kahneman knew how to improve the Israeli army interviewing system. His new system did not allow the interviewers the luxury of free-ranging conversations. Instead, they were required to ask a standard list of factual questions about the life and work of each recruit. The answers were then converted into numerical scores, and the scores were inserted into formulas measuring the aptitude of the recruit for the various army jobs. When the predictions of the new system were compared to performances several months later, the results showed the new system to be much better than the old. Statistics and simple arithmetic tell us more about ourselves than expert intuition.

Reflecting fifty years later on his experience in the Israeli army, Kahneman remarks in *Thinking, Fast and Slow** that it was not unusual in those days for young people to be given big responsibilities.

---

*Farrar, Straus and Giroux, 2011.

The country itself was only seven years old. "All its institutions were under construction," he says, "and someone had to build them." He was lucky to be given this chance to share in the building of a country, and at the same time to achieve an intellectual insight into human nature. He understood that the failure of the old interview system was a special case of a general phenomenon that he called "the *illusion of validity*." At this point, he says, "I had discovered my first cognitive illusion."

Cognitive illusions are the main theme of his book. A cognitive illusion is a false belief that we intuitively accept as true. The illusion of validity is a false belief in the reliability of our own judgment. The interviewers sincerely believed that they could predict the performance of recruits after talking with them for fifteen minutes. Even after the interviewers had seen the statistical evidence that their belief was an illusion, they still could not help believing it. Kahneman confesses that he himself still experiences the illusion of validity, after fifty years of warning other people against it. He cannot escape the illusion that his own intuitive judgments are trustworthy.

An episode from my own past is curiously similar to Kahneman's experience in the Israeli army. I was a statistician before I became a scientist. At the age of twenty I was doing statistical analysis of the operations of the British Bomber Command in World War II. The command was then seven years old, like the State of Israel in 1955. All its institutions were under construction. It consisted of six bomber groups that were evolving toward operational autonomy. Air Vice-Marshal Sir Ralph Cochrane was the commander of 5 Group, the most independent and effective of the groups. Our bombers were then taking heavy losses, the main cause of loss being the German night fighters.

Cochrane said the bombers were too slow, and the reason they were too slow was that they carried heavy gun turrets that increased their aerodynamic drag and lowered their operational ceiling. Because the bombers flew at night, they were normally painted black. Being a flamboyant character, Cochrane announced that he would like to take a Lancaster bomber, rip out the gun turrets and all the associated dead weight, ground the two gunners, and paint the whole thing white. Then he would fly it over Germany, and fly so high and so fast that nobody could shoot him down. Our commander in chief did not approve of this suggestion, and the white Lancaster never flew.

The reason why our commander in chief was unwilling to rip out gun turrets, even on an experimental basis, was that he was blinded by the illusion of validity. This was ten years before Kahneman discovered it and gave it its name, but the illusion of validity was already doing its deadly work. All of us at Bomber Command shared the illusion. We saw every bomber crew as a tightly knit team of seven, with the gunners playing an essential role defending their comrades against fighter attack, while the pilot flew an irregular corkscrew to defend them against flak. An essential part of the illusion was the belief that the team learned by experience. As they became more skillful and more closely bonded, their chances of survival would improve.

When I was collecting the data in the spring of 1944, the chance of a crew reaching the end of a thirty-operation tour was about 25 percent. The illusion that experience would help them to survive was essential to their morale. After all, they could see in every squadron a few revered and experienced old-timer crews who had completed one tour and had volunteered to return for a second tour. It was obvious to everyone that the old-timers survived because they were more skillful. Nobody wanted to believe that the old-timers survived only because they were lucky.

At the time Cochrane made his suggestion of flying the white Lancaster, I had the job of examining the statistics of bomber losses. I did a careful analysis of the correlation between the experience of the crews and their loss rates, subdividing the data into many small packages so as to eliminate effects of weather and geography. My results were as conclusive as those of Kahneman. There was no effect of experience on loss rate. So far as I could tell, whether a crew lived or died was purely a matter of chance. Their belief in the lifesaving effect of experience was an illusion.

The demonstration that experience had no effect on losses should have given powerful support to Cochrane's idea of ripping out the gun turrets. But nothing of the kind happened. As Kahneman found out later, the illusion of validity does not disappear just because facts prove it to be false. Everyone at Bomber Command, from the commander in chief to the flying crews, continued to believe in the illusion. The crews continued to die, experienced and inexperienced alike, until Germany was overrun and the war finally ended.

Another theme of Kahneman's book, proclaimed in the title, is the existence in our brains of two independent systems for organizing knowledge. Kahneman calls them System 1 and System 2. System 1 is amazingly fast, allowing us to recognize faces and understand speech in a fraction of a second. It must have evolved from the ancient little brains that allowed our agile mammalian ancestors to survive in a world of big reptilian predators. Survival in the jungle requires a brain that makes quick decisions based on limited information. Intuition is the name we give to judgments based on the quick action of System 1. It makes judgments and takes action without waiting for our conscious awareness to catch up with it. The most remarkable fact about System 1 is that it has immediate access to a vast store of memories that it uses as a basis for judgment. The memories that are most accessible are those associated with strong emotions, with fear and pain and hatred. The resulting judgments are often wrong, but in the world of the jungle it is safer to be wrong and quick than to be right and slow.

System 2 is the slow process of forming judgments based on conscious thinking and critical examination of evidence. It appraises the actions of System 1. It gives us a chance to correct mistakes and revise opinions. It probably evolved more recently than System 1, after our primate ancestors became arboreal and had the leisure to think things over. An ape in a tree is not so much concerned with predators as with the acquisition and defense of territory. System 2 enables a family group to make plans and coordinate activities. After we became human, System 2 enabled us to create art and culture.

The question then arises: Why do we not abandon the error-prone System 1 and let the more reliable System 2 rule our lives? Kahneman gives a simple answer to this question: System 2 is lazy. To activate System 2 requires mental effort. Mental effort is costly in time and also in calories. Precise measurements of blood chemistry show that consumption of glucose increases when System 2 is active. Thinking is hard work, and our daily lives are organized so as to economize on thinking. Many of our intellectual tools, such as mathematics and rhetoric and logic, are convenient substitutes for thinking. So long as we are engaged in the routine skills of calculating and talking and writing, we are not thinking, and System 1 is in charge. We only make the mental effort to activate System 2 after we have exhausted the possible alternatives.

System 1 is much more vulnerable to illusions, but System 2 is not immune to them. Kahneman uses the phrase "availability bias" to mean a biased judgment based on a memory that happens to be quickly available. It does not wait to examine a bigger sample of less cogent memories. A striking example of availability bias is the fact that sharks save the lives of swimmers. Careful analysis of deaths in the ocean near San Diego shows that on average, the death of each swimmer killed by a shark saves the lives of ten others. Every time a swimmer is killed, the number of deaths by drowning goes down for a few years and then returns to the normal level. The effect occurs because reports of death by shark attack are remembered more vividly than reports of drownings. System 1 is strongly biased, paying more prompt attention to sharks than to riptides that occur more frequently and may be equally lethal. In this case, System 2 probably shares the same bias. Memories of shark attacks are tied to strong emotions and are therefore more available to both systems.

Kahneman is a psychologist who won a Nobel Prize in Economics. His great achievement was to turn psychology into a quantitative science. He made our mental processes subject to precise measurement and exact calculation, by studying in detail how we deal with dollars and cents. By making psychology quantitative, he incidentally achieved a powerful new understanding of economics. A large part of his book is devoted to stories illustrating the various illusions to which supposedly rational people succumb. Each story describes an experiment, examining the behavior of students or citizens who are confronted with choices under controlled conditions. The subjects make decisions that can be precisely measured and recorded. The majority of the decisions are numerical, concerned with payments of money or calculations of probability. The stories demonstrate how far our behavior differs from the behavior of the mythical "rational actor" who obeys the rules of classical economics.

A typical example of a Kahneman experiment is the coffee mug experiment, designed to measure a form of bias that he calls the "endowment effect." The endowment effect is our tendency to value an object more highly when we own it than when someone else owns it. Coffee mugs are intended to be useful as well as elegant, so that people who own them become personally attached to them. A simple version of the experiment has two groups of people, sellers and buyers, picked at random from a population of students. Each seller is given a mug and invited to sell it to a buyer. The buyers are given nothing and are invited to use their own money to buy a mug from a seller. The average prices offered in a typical experiment were: sellers $7.12, buyers $2.87. Because the price gap was so large, few mugs were actually sold.

The experiment convincingly demolished the central dogma of classical economics. The central dogma says that in a free market, buyers and sellers will agree on a price that both sides regard as fair. The dogma is true for professional traders trading stocks in a stock market. It is untrue for nonprofessional buyers and sellers because of the endowment effect. Trading that should be profitable to both sides does not occur, because most people do not think like traders.

Our failure to think like traders has important practical consequences, for good and for evil. The main consequence of the endowment effect is to give stability to our lives and institutions. Stability is good when a society is peaceful and prosperous. Stability is evil when a society is poor and oppressed. The endowment effect works for good in the German city of Munich. I once rented a home there for a year, a few miles from the city center. Across the street from our home was a real farm with potato fields and pigs and sheep. The local children, including ours, went out to the fields after dark, made little fires in the ground, and roasted potatoes. In a free-market economy, the farm would have been sold to a developer and converted into a housing development. The farmer and the developer would both have made a handsome profit. But in Munich, people were not thinking like traders. There was no free market in land. The city valued the farm as public open space, allowing city dwellers to walk over grass all the way to the city center, and allowing our children to roast potatoes at night. The endowment effect allowed the farm to survive.

In poor agrarian societies, such as Ireland in the nineteenth century or much of Africa today, the endowment effect works for evil because it perpetuates poverty. For the Irish landowner and the African village chief, possessions bring status and political power. They do not think like traders, because status and political power are more valuable than money. They will not trade their superior status for money, even when they are heavily in debt. The endowment effect keeps the peasants poor and drives those of them who think like traders to emigrate.

At the end of his book, Kahneman asks the question: What practical benefit can we derive from an understanding of our irrational mental processes? We know that our judgments are heavily biased by inherited illusions, which helped us to survive in a snake-infested jungle but have nothing to do with logic. We also know that, even when we become aware of the bias and the illusions, the illusions do not disappear. What use is it to know that we are deluded, if the knowledge does not dispel the delusions?

Kahneman answers this question by saying that he hopes to change our behavior by changing our vocabulary. If the names that he invented for various common biases and illusions, "illusion of validity," "availability bias," "endowment effect," and others that I have no space to describe here, become part of our everyday vocabulary, then he hopes to see the illusions lose their power to deceive us. If we use these names every day to criticize our friends' mistaken judgments and to confess our own, then perhaps we will learn to overcome our illusions. Perhaps our children and grandchildren will grow up using the new vocabulary and will automatically correct their congenital biases when making judgments. If this miracle happens, then future generations will owe a big debt to Kahneman for giving them a clearer vision.

One thing that is notably absent from Kahneman's book is the name of Sigmund Freud. In thirty-two pages of endnotes there is not a single reference to his writings. This omission is certainly no accident. Freud was a dominating figure in the field of psychology for the first half of the twentieth century, and a fallen tyrant for the second half of the century. In the article on Freud in Wikipedia, we find quotes from the Nobel Prize–winning immunologist Peter Medawar —psychoanalysis is the "most stupendous intellectual confidence trick of the twentieth century"—and from Frederick Crews:

> Step by step, we are learning that Freud has been the most overrated figure in the entire history of science and medicine—one who wrought immense harm through the propagation of false etiologies, mistaken diagnoses, and fruitless lines of enquiry.

In these quotes, emotions are running high. Freud is now hated as passionately as he was once loved. Kahneman evidently shares the prevalent repudiation of Freud and of his legacy of writings.

Freud wrote two books, *The Psychopathology of Everyday Life* in 1901 and *The Ego and the Id* in 1923, which come close to preempting two of the main themes of Kahneman's book. The psychopathology book describes the many mistakes of judgment and of action that arise from emotional bias operating below the level of consciousness. These "Freudian slips" are examples of availability bias, caused by memories associated with strong emotions. *The Ego and the Id* describes two levels of the mind that are similar to the System 2 and System 1 of Kahneman, the ego being usually conscious and rational, the id usually unconscious and irrational.

There are huge differences between Freud and Kahneman, as one would expect for thinkers separated by a century. The deepest difference is that Freud is literary while Kahneman is scientific. The great contribution of Kahneman was to make psychology an experimental science, with experimental results that could be repeated and verified. Freud, in my view, made psychology a branch of literature, with stories and myths that appeal to the heart rather than to the mind. The central dogma of Freudian psychology was the Oedipus complex, a story borrowed from Greek mythology and enacted in the tragedies of Sophocles. Freud claimed that he had identified from his clinical practice the emotions children feel toward their parents that he called the Oedipus complex. His critics have rejected that claim. So Freud became to his admirers a prophet of spiritual and psychological wisdom, and to his detractors a quack doctor pretending to cure imaginary diseases. Kahneman took psychology in a diametrically opposite direction, not pretending to cure ailments but only trying to dispel illusions.

It is understandable that Kahneman has no use for Freud, but it is still regrettable. The insights of Kahneman and Freud are complementary rather than contradictory. Anyone who strives for a complete understanding of human nature has much to learn from both of them. The scope of Kahneman's psychology is necessarily limited by his method, which is to study mental processes that can be observed and measured under rigorously controlled experimental conditions. Following this method, he revolutionized psychology. He discovered mental processes that can be described precisely and demonstrated reliably. He discarded the poetic fantasies of Freud.

But along with the poetic fantasies, he discarded much else that was valuable. Since strong emotions and obsessions cannot be experimentally controlled, Kahneman's method did not allow him to study them. The part of the human personality that Kahneman's method can handle is the nonviolent part, concerned with everyday decisions, artificial parlor games, and gambling for small stakes. The violent and passionate manifestations of human nature, concerned with matters of life and death and love and hate and pain and sex, cannot be experimentally controlled and are beyond Kahneman's reach. Violence and passion are the territory of Freud. Freud can penetrate deeper than Kahneman because literature digs deeper than science into human nature and human destiny.

William James is another great psychologist whose name is not mentioned in Kahneman's book. James was a contemporary of Freud and published his classic work, *The Varieties of Religious Experience: A Study in Human Nature*, in 1902. Religion is another large area of human behavior that Kahneman chooses to ignore. Like the Oedipus complex, religion does not lend itself to experimental study. Instead of doing experiments, James listens to people describing their experiences. He studies the minds of his witnesses from the inside rather than from the outside. He finds the religious temperament divided into two types that he calls once-born and twice-born, anticipating Kahneman's division of our minds into System 1 and System 2. Since James turns to literature rather than to science for his evidence, the two chief witnesses that he examines are Walt Whitman for the once-born and Leo Tolstoy for the twice-born.

Freud and James were artists and not scientists. It is normal for artists who achieve great acclaim during their lifetimes to go into eclipse and become unfashionable after their deaths. Fifty or a hundred years later, they may enjoy a revival of their reputations, and they may then be admitted to the ranks of permanent greatness. Admirers of Freud and James may hope that the time may come when they will stand together with Kahneman as three great explorers of the human psyche, Freud and James as explorers of our deeper emotions, Kahneman as the explorer of our more humdrum cognitive processes. But that time has not yet come. Meanwhile, we must be grateful to Kahneman for giving us in this book a joyful understanding of the practical side of our personalities.

Note added in 2014: Kahneman responded to the review in a letter to the editor:

> Freeman Dyson's generous review... greatly overstates my role in the story of scientific psychology. My discipline is indeed much more scientific than it was when William James and Sigmund Freud wrote their masterpieces, but the transformation was well underway long before I was born. The science of psychology grew in several stages in the twentieth century, from the schools of Gestalt psychology and behaviorism that I learned about as a graduate student around 1960, on to the cognitive revolution that was reshaping the intellectual landscape when Amos Tversky and I began our collaboration at the end of that decade, and from there to the developments in neuroscience and in the study of associative and emotional processes that attract many of the best graduate students of today.
>
> Tversky and I were participants in the cognitive revolution, to which we initially contributed the idea that significant errors of intuitive judgment can arise from the mechanism of cognition, rather than from wishful thinking or other emotional distortions. We also had a glimmer of what later became the two-system idea. Our first joint paper, which documented mistakes in the statistical decisions of researchers, informally distinguished intuitive judgment from deliberate computation. The detailed study of contrasts between automatic and controlled processes began some years later in an Indiana laboratory, and many psychologists have refined and extended that distinction in the intervening decades. In my recent attempt to describe the interactions between fast intuitive thinking and

the deliberate self, I draw both on the work of these predecessors and on recent advances in the study of associative memory.

Scientists operate mostly in disciplinary silos, and it is rare for research in one field to influence work in other disciplines. My research with Tversky crossed some of these boundaries, in large part because of our use of demonstrations that were accessible to everyone: we engaged readers in simple problems in which they could observe errors in their own intuitions. Our work has therefore been more visible to outsiders than many other advances in psychological research, but it is best seen as a contribution to the thriving collective enterprise of modern experimental psychology.

This letter does not respond to my suggestion that psychology lives on both sides of the boundary between science and literature, combining the emotional insights of Freud and James with the experimental discoveries of Tversky and Kahneman. As I said in the review, the time for reconciliation has not yet come.

17

WHAT CAN YOU REALLY KNOW?

JIM HOLT'S *Why Does the World Exist?: An Existential Detective Story* is a portrait gallery of leading modern philosophers.* He visited each of them in turn, warning them in advance that he was coming to discuss with them a single question: Why is there something rather than nothing? He reports their reactions to this question, and embellishes their words with descriptions of their habits and personalities. Their answers give us vivid glimpses of the speakers but do not solve the riddle of existence.

The philosophers are more interesting than the philosophy. Most of them are eccentric characters who have risen to the top of their profession. They think their deep thoughts in places of unusual beauty such as Paris and Oxford. They are heirs to an ancient tradition of academic hierarchy, in which disciples sat at the feet of sages, and sages enlightened disciples with Delphic utterances. The universities of Paris and Oxford have maintained this tradition for eight hundred years. The great world religions have maintained it even longer. Universities and religions are the most durable of human institutions.

*Liveright, 2012.

According to Holt, the two most influential philosophers of the twentieth century were Martin Heidegger and Ludwig Wittgenstein: Heidegger supreme in continental Europe, Wittgenstein in the English-speaking world. Heidegger was one of the founders of existentialism, a school of philosophy that was especially attractive to French intellectuals. Heidegger himself lost his credibility in 1933 when he accepted the position of rector of the University of Freiburg under the newly established Hitler government and became a member of the Nazi Party. Existentialism continued to flourish in France after it faded in Germany.

Wittgenstein, unlike Heidegger, did not establish an ism. He wrote very little, and everything that he wrote was simple and clear. The only book that he published during his lifetime was *Tractatus Logico-Philosophicus*, written in Vienna in 1918 and published in England with a long introduction by Bertrand Russell in 1922. It fills less than two hundred small pages, even though the original German and the English translation are printed side by side. I was lucky to be given a copy of the *Tractatus* as a prize when I was in high school. I read it through in one night, in an ecstasy of adolescent enthusiasm. Most of it is about mathematical logic. Only the last five pages deal with human problems. The text is divided into numbered sections, each consisting of one or two sentences. For example, section 6.521 says: "The solution of the problem of life is seen in the vanishing of this problem. Is not this the reason why men, to whom after long doubting the sense of life became clear, could not then say wherein this sense consisted?" The most famous sentence in the book is the final section 7: "Wherof one cannot speak, thereof one must be silent."

I found the book enlightening and liberating. It said that philosophy is simple and has limited scope. Philosophy is concerned with logic and the correct use of language. All speculations outside this limited area are mysticism. Section 6.522 says: "There is indeed the inexpressible. This shows itself. It is the mystical." Since the mystical is inexpressible, there is nothing more to be said. Holt summarizes the difference between Heidegger and Wittgenstein in nine words: "Wittgenstein was brave and ascetic, Heidegger treacherous and vain." These words apply equally to their characters as human beings and to their intellectual output.

Wittgenstein's intellectual asceticism had a great influence on the philosophers of the English-speaking world. It narrowed the scope of philosophy by excluding ethics and aesthetics. At the same time, his personal asceticism enhanced his credibility. During World War II, he wanted to serve his adopted country in a practical way. Being too old for military service, he took a leave of absence from his academic position in Cambridge and served in a menial job, as a hospital orderly taking care of patients. When I arrived at Cambridge University in 1946, Wittgenstein had just returned from his six years of duty at the hospital. I held him in the highest respect and was delighted to find him living in a room above mine on the same staircase. I frequently met him walking up or down the stairs, but I was too shy to start a conversation. Several times I heard him muttering to himself: "I get stupider and stupider every day."

Finally, toward the end of my time in Cambridge, I ventured to speak to him. I told him I had enjoyed reading the *Tractatus*, and I asked him whether he still held the same views that he had expressed twenty-eight years earlier. He remained silent for a long time and then said, "Which newspaper do you represent?" I told him I was a student and not a journalist, but he never answered my question.

Wittgenstein's response to me was humiliating, and his response to female students who tried to attend his lectures was even worse. If a woman appeared in the audience, he would remain standing silent until she left the room. I decided that he was a charlatan using outrageous behavior to attract attention. I hated him for his rudeness.

Fifty years later, walking through a churchyard on the outskirts of Cambridge on a sunny morning in winter, I came by chance upon his tombstone, a massive block of stone lightly covered with fresh snow. On the stone was written a single word: WITTGENSTEIN. To my surprise, I found that the old hatred was gone, replaced by a deeper understanding. He was at peace, and I was at peace too, in the white silence. He was no longer an ill-tempered charlatan. He was a tortured soul, the last survivor of a family with a tragic history, living a lonely life among strangers, trying until the end to express the inexpressible.

The philosophers that Holt interviewed wander over a wide landscape. The main theme of their discussions is a disagreement between two groups that I call materialists and Platonists. Materialists imagine a world built out of atoms. Platonists imagine a world built out of ideas. This division into two categories is a gross simplification, lumping together people with a great variety of opinions. Like taxonomists who name species of plants and animals, observers of the philosophical scene may be splitters or lumpers. Splitters like to name many species; lumpers like to name few.

Holt is a splitter and I am a lumper. Philosophers are mostly splitters, dividing their ways of thinking into narrow specialties such as theism or deism or humanism or panpsychism or axiarchism. Examples of each of these isms are to be seen in Holt's collection. I find it more convenient to lump them into two big groups: one obsessed with matter and the other obsessed with mind. Holt asks them to explain why the world exists. For the materialists, the question concerns the origin of space and time and particles and fields, and the relevant branch of science is physics. For the Platonists, the question concerns the origin of meaning and purpose and consciousness, and the relevant science is psychology.

The most impressive of the Platonists is John Leslie, who spent most of his life teaching philosophy at the University of Guelph and is now living in retirement on the west coast of Canada. He calls himself an extreme axiarchist. The word "axiarchism" is Greek for "value rules," meaning that the world is built out of ideas, and the Platonic idea of the Good gives value to everything that exists. Leslie takes seriously Plato's image of the cave as a metaphor of human life. We live in a cave, seeing only shadows cast on the wall by light streaming in from the entrance. The real objects outside the cave are ideas, and all the things that we perceive inside are imperfect images of ideas. Evil exists because our images are distorted. The ultimate reality hidden from our view is Goodness. Goodness is a strong enough force to pull the universe into existence. Leslie understands that this explanation of existence is a poetic fantasy rather than a logical argument. Fantasy comes to the rescue when logic fails. The whole range of Plato's thinking is embodied in his dialogues, which are dramatic reconstructions of the conversations of his master Socrates. They are based on imagination, not on logic.

In 1996 Leslie published a book, *The End of the World*, taking a gloomy view of the human situation. He was calculating the probable future duration of the human species, basing his argument on the Copernican principle, which says that the situation of the human observer in the cosmos should be in no way exceptional. Copernicus gave his name to this principle when he moved the earth from its position at the center of the Aristotelian universe and put it into a more modest position as one of the planets orbiting around the sun.

Leslie argued that the Copernican principle should apply to our position in time as well as to our position in space. As observers of the passage of time, we should not put ourselves into a privileged position at the beginning of the history of our species. As Copernican observers, we should expect to be in an average position in our history, rather than close to the beginning. Therefore, we should expect the future duration of our species to be not much longer than its past. Since we know that our species originated about a hundred thousand years ago, we should expect it to become extinct about a hundred thousand years from now.

When Leslie published this prognostication, I protested strongly against it, claiming that it was a technically wrong use of the theory of probability. In fact Leslie's argument was technically correct. The reason I did not like the argument was that I did not like the conclusion. I thought that the universe had a purpose, and that our minds were a part of that purpose. Since the goodness of the universe was revealed in our existence as observers, we could rely on the goodness of the universe to allow us to continue to exist. I opposed Leslie's argument because I was a better Platonist than he was.

The antithesis of John Leslie is David Deutsch, whose book *The Beginning of Infinity* I discuss in chapter 12. Holt visited Deutsch at his home in a village a few miles from Oxford. The chapter describing the visit is entitled "The Magus of the Multiverse." Deutsch is a professional physicist who uses physics as a basis for philosophical speculation. Unlike most philosophers, he understands quantum mechanics and feels at home in a quantum universe. He likes the many-universe interpretation of quantum mechanics, invented in the 1950s by Hugh Everett, who was then a student in Princeton. Everett imagined the quantum universe as an infinite assemblage of ordinary universes all existing simultaneously. He called the assemblage "the multiverse."

The essence of quantum physics is unpredictability. At every instant, the objects in our physical environment—the atoms in our lungs and the light in our eyes—are making unpredictable choices, deciding what to do next. According to Everett and Deutsch, the multiverse contains a universe for every combination of choices. There are so many universes that every possible sequence of choices occurs in at least one of them. Each universe is constantly splitting into many alternative universes, and the alternatives are recombining when they arrive at the same final state by different routes. The multiverse is a huge network of possible histories diverging and reconverging as time goes on. The "quantum weirdness" that we observe in the behavior of atoms, the "spooky action at a distance" that Einstein famously disliked, is the result of universes recombining in unexpected ways.

According to Deutsch, each of us exists in the multiverse as a crowd of almost identical creatures, traveling together through time along closely related histories, splitting and recombining constantly like the atoms of which we are composed. He does not claim to have an answer to the question "Why does the multiverse exist?" or to the easier question "What is the nature of consciousness?" He sees ahead of us a long future of slow exploration, answering philosophical questions that we do not yet know how to ask. One of the questions that we know how to ask but not to answer is "Does quantum computing play an essential role in our consciousness?" For Deutsch, the physics of quantum computing is the most promising clue that may lead us to a deeper understanding of our existence. He theorizes that all the different parallel universes in the multiverse could be coaxed into collaborating on a single computation.

There are many other kinds of multiverse besides the Everett version. Multiverse models are fashionable in recent theories of cosmology. Holt went to see the Russian cosmologist Alex Vilenkin at Tufts University in Boston. Unlike Deutsch, Vilenkin has multiple universes disconnected and widely separated from each other. Each arises out of nothing by a process known as quantum tunneling, spontaneously crossing the barrier between nonexistence and existence with no expenditure of energy. Universes spring into existence with precisely zero total energy, the positive energy of matter being equal and opposite to the negative energy of gravitation. Mass comes free because energy is zero.

The title of the Vilenkin chapter is "The Ultimate Free Lunch?" Holt describes a conversation between the young physicist George Gamow and the old physicist Albert Einstein when both of them were in Princeton. Gamow, the original inventor of the idea of quantum tunneling, explained to Einstein the possibility of the free lunch. Einstein was so astonished that he stopped in the middle of the street and was almost run over by a car.

Opinions vary widely concerning the proper limits of science. For me, the multiverse is philosophy and not science. Science is about facts that can be tested and mysteries that can be explored, and I see no way of testing hypotheses of the multiverse. Philosophy is about ideas that can be imagined and stories that can be told. I put narrow limits on science, but I recognize other sources of human wisdom going beyond science. Other sources of wisdom are literature, art, history, religion, and philosophy. The multiverse has its place in philosophy and in literature.

My favorite version of the multiverse is a story told by the philosopher Olaf Stapledon, who died in 1950. He taught philosophy at the University of Liverpool. In 1937 he published a novel, *Star Maker*, describing his vision of the multiverse. The book was marketed as science fiction, but it has more to do with theology than with science. The narrator has a vision in which he travels through space visiting alien civilizations from the past and the future, his mind merging telepathically with some of their inhabitants who join him on his journey. Finally, this "cosmical mind" encounters the Star Maker, an "eternal and absolute spirit" who has created all of these worlds in a succession of experiments. Each experiment is a universe, and as each experiment fails he learns how to design the next experiment a little better. His first experiment is a simple piece of music, a rhythmic drumbeat exploring the texture of time. After that come many more works of art, exploring the possibilities of space and time with gradually increasing complexity.

Our own universe comes somewhere in the middle, a big improvement on its predecessors but still destined for failure. Its flaws will bring it to a tragic end. Far outside the range of our understanding will be the later experiments, avoiding the mistakes that the Star Maker made with our own universe and leading the way to ultimate perfection. Stapledon's multiverse, conceived in the shadow of the approaching horrors of World War II, is an imaginative attempt to grapple with the problem of good and evil.

For most of the twenty-five centuries since written history began, philosophers were important. Two groups of philosophers, Confucius and Lao Tse in China and Socrates, Plato, and Aristotle in Greece, were dominant figures in the cultures of Asia and Europe for two thousand years. Confucius and Aristotle set the style of thinking for Eastern and Western civilizations. They spoke not only to scholars but also to rulers. They had a deep influence in the practical worlds of politics and morality as well as in the intellectual worlds of science and scholarship.

In more recent centuries, philosophers were still leaders of human destiny. Descartes and Montesquieu in France, Spinoza in Holland, Hobbes and Locke in England, Hegel and Nietzsche in Germany, set their stamp on the divergent styles of nations as nationalism became the driving force in the history of Europe. Through all the vicissitudes of history, from classical Greece and China until the end of the nineteenth century, philosophers were giants playing a dominant role in the kingdom of the mind.

Holt's philosophers belong to the twentieth and twenty-first centuries. Compared with the giants of the past, they are a sorry bunch of dwarfs. They are thinking deep thoughts and giving scholarly lectures to academic audiences, but hardly anybody in the world outside is listening. They are historically insignificant. At some time toward the end of the nineteenth century, philosophers faded from public life. Like the baker in Lewis Carroll's poem "The Hunting of the Snark," they suddenly and silently vanished. So far as the general public was concerned, philosophers became invisible.

The fading of philosophy came to my attention in 1979, when I was involved in the planning of a conference to celebrate the hundredth birthday of Einstein. The conference was held in Princeton, where Einstein had lived, and our largest meeting hall was too small for all the people who wanted to come. A committee was set up to decide who should be invited. When the membership of the committee was announced, there were loud protests from people who were excluded. After acrimonious discussions, we agreed to have three committees, each empowered to invite one third of the participants. One committee was for scientists, one for historians of science, and one for philosophers of science.

After the three committees had made their selections, we had three lists of names of people to be invited. I looked at the lists of names and was immediately struck by their disconnection. With a few exceptions, I knew personally all the people on the science list. On the history list, I knew the names, but I did not know the people personally. On the philosophy list, I did not even know the names.

In earlier centuries, scientists and historians and philosophers would have known one another. Newton and Locke were friends and colleagues in the English Parliament of 1689, helping to establish constitutional government in England after the bloodless revolution of 1688. The bloody passions of the English civil war were finally quieted by establishing a constitutional monarchy with limited powers. Constitutional monarchy was a system of government invented by philosophers. But in the twentieth century, science and history and philosophy had become separate cultures. We were three groups of specialists, living in separate communities and rarely speaking to one another.

When and why did philosophy lose its bite? How did it become a toothless relic of past glories? These are the ugly questions that Holt's book compels us to ask. Philosophers became insignificant when philosophy became a separate academic discipline, distinct from science and history and literature and religion. The great philosophers of the past covered all these disciplines. Until the nineteenth century, science was called natural philosophy and officially recognized as a branch of philosophy. The word "scientist" was invented by William Whewell, a nineteenth-century Cambridge philosopher who became master of Trinity College and put his name on the building where Wittgenstein and I were living in 1946. Whewell introduced the word in 1833. He was waging a deliberate campaign to establish science as a professional discipline distinct from philosophy.

Whewell's campaign succeeded. As a result, science grew to a dominant position in public life and philosophy shrank. Philosophy shrank even further when it became detached from religion and from literature. The great philosophers of the past wrote literary masterpieces such as the book of Job and the *Confessions* of Saint Augustine. The latest masterpieces written by a philosopher were probably Friedrich Nietzsche's *Thus Spoke Zarathustra* in 1885 and *Beyond Good and Evil* in 1886. Modern departments of philosophy have no place for the mystical.

Note added in 2014: In the published review, I had the snark vanishing instead of the baker. Thanks to Ray Fair for correcting this grievous error. Thanks to Graham Farmelo, the author of a book about Winston Churchill that I reviewed later (see chapter 20), for telling me that Churchill was a fan of Olaf Stapledon, eagerly reading everything that Stapledon wrote.

18

OPPENHEIMER: THE SHAPE OF GENIUS

WHY ANOTHER BOOK about Robert Oppenheimer? Many books have been written and widely read, ranging from the impressionistic *Lawrence and Oppenheimer* of Nuel Pharr Davis to the scholarly *American Prometheus* of Kai Bird and Martin Sherwin. Ray Monk says he wrote *Robert Oppenheimer: A Life Inside the Center** because the others gave too much weight to Oppenheimer's politics and too little weight to his science. Monk restores the balance by describing in detail the activities that occupied most of Oppenheimer's life: learning and exploring and teaching science.

The subtitle, "A Life Inside the Center," calls attention to a rarer skill at which Oppenheimer excelled. He had a unique ability to put himself at the places and times at which important things were happening. Four times in his life, he was at the center of important events. In 1926 he was at Göttingen, where his teacher Max Born was one of the leaders of the quantum revolution that transformed our view of the subatomic world. In 1929 he was at Berkeley, where his friend Ernest Lawrence was building the first cyclotron, and with Lawrence he created in Berkeley an American school of subatomic

*Doubleday, 2013.

physics that took the leadership away from Europe. In 1943 he was in Los Alamos building the first nuclear weapons. In 1947 he was in Washington as the chairman of the General Advisory Committee of the United States Atomic Energy Commission, giving advice to political and military leaders at the highest levels of government. He was driven by an irresistible ambition to play a leading part in historic events. In each case, when he was present at the center of action, he rose to the occasion and took charge of the situation with unexpected competence.

It is often helpful to have several books covering the same territory. Since different writers have different viewpoints, each book will do better in some areas and worse in others. The most valuable contribution of Monk's book is to give a detailed picture of two groups of people who played an important role in Oppenheimer's life: the tightly knit society of wealthy German New York Jews to which his parents belonged, and the small army of security officers who monitored his social and political activities when he was engaged in secret work in Berkeley and Los Alamos.

Monk brings these two groups vividly to life. He puts the German Jews into their historical setting. Many of them were liberal idealists who failed to achieve their dreams of social reform in Germany and came to America with an intense commitment to the American dream of a free society. He begins his account by quoting a line, "America, thou hast it better," from a poem by Goethe extolling America as the land of liberation from the knights, robbers, and ghosts of old Europe. This poetic German vision of America made Oppenheimer more passionately patriotic than most of his scientist friends. His father was a close friend of Felix Adler, the founder of the Ethical Culture Society, an institution that embodied the liberal ideals of the German Jewish community. Oppenheimer was educated at the Ethical Culture School, also founded by Felix Adler, from age seven to age seventeen. The purpose of the school was to raise children with a lifelong dedication to ethical principles, detached from any religious belief or ritual. The school seems to have been successful in molding Oppenheimer's character.

Unlike the other books, which portray the security officers at Berkeley and Los Alamos as mindless bureaucrats or paranoid witch-hunters, Monk portrays them as real people with real problems. The four officers who had the thankless job of extracting information from Oppenheimer were Boris Pash, Peer de Silva, John Lansdale, and Lyall Johnson. They were trying conscientiously to protect the secrecy of the bomb project and to keep potential spies out of it.

We now know that their efforts were unsuccessful. They did not succeed in identifying the real spies. But they were aware that Soviet intelligence agents were actively seeking information about the project; they suspected that several of Oppenheimer's Communist friends and students might be spying; and they were frustrated by Oppenheimer's evasive answers to their questions. They were told by General Leslie Groves, the commander in chief of the bomb project, that Oppenheimer's leadership was essential to it, and yet Oppenheimer disregarded the security rules that they were trying to enforce. In their eyes, the essential question was whether the security rules that applied to everyone else on the project should also apply to Oppenheimer. Should he be exempt from the rules just because he was famous? Lansdale answered yes to this question. Pash, de Silva, and Johnson answered no.

Monk begins his book with a preface discussing the other Oppenheimer books and explaining why he finds them deficient. Unfortunately he does not mention the book that I found the most illuminating, *Reappraising Oppenheimer: Centennial Studies and*

Reflections, edited by Cathryn Carson and David Hollinger.* This work does not appear in Monk's bibliography. It is a volume of essays by various authors, most of them professional historians. I summarize here three of the essays that provide factual information about aspects of Oppenheimer's life that Monk does not explore.

David Cassidy gives us a complete list of Oppenheimer's graduate students with the titles of their dissertations. This list shows us what Oppenheimer was thinking about when he was a young professor in Berkeley, and how he was training the next generation of physicists. It reveals the modest foundation out of which his legendary reputation as a teacher grew. There were twenty-five students altogether, only six of them finishing in the ten years between 1929 and 1939. Two subjects are predominant in their dissertations: cosmic rays and mesons. Cosmic rays are the gentle rain of high-energy particles that constantly bombard the earth from outer space. Mesons are particles that were discovered among the cosmic rays and found to have weird and incomprehensible behavior, sometimes exploding into showers of secondary particles, and sometimes passing through matter without interaction.

Oppenheimer knew that the cosmic rays were his best clue to the understanding of nature in the wild, with energies vastly greater than Lawrence could reach with his particle accelerators. A majority of his students worked on cosmic rays in one way or another, trying to understand a mass of confusing observations by comparing them with a mass of equally confusing theories. This intense intellectual effort succeeded in educating gifted students who went on to become scientific leaders, but it failed to solve the mysteries of cosmic rays and mesons.

---

*It is volume 21 in the series Berkeley Papers in History of Science, published in 2005 by the University of California.

David Holloway gives us a chapter entitled "Parallel Lives? Oppenheimer and Khariton," comparing Oppenheimer with his Russian counterpart. Yulii Khariton was remarkably similar to Oppenheimer in many ways, born in the same year into a cultured Jewish family, fluent in three languages, with a strong interest in art and literature, working as a student at the Cavendish Laboratory in England just after Oppenheimer left, and unexpectedly becoming the successful leader of a Soviet program to build bombs. His boss, Lavrenty Beria, was a notoriously harsh KGB official, but Khariton succeeded in working with Beria as harmoniously as Oppenheimer did with General Groves. Khariton never became a controversial public figure like Oppenheimer. His close friend and colleague Andrei Sakharov, who was for many years his deputy, played that part in the Soviet Union.

Karl Hufbauer contributes a chapter, "J. Robert Oppenheimer's Path to Black Holes," discussing what I consider the outstanding mystery of Oppenheimer's life. In 1939 Oppenheimer published with his student Hartland Snyder the paper "On Continued Gravitational Contraction," only four pages long, which is in my opinion Oppenheimer's one and only revolutionary contribution to science. In that paper, Oppenheimer and Snyder invented the concept of black holes; they proved that every star significantly more massive than the sun must end its life as a black hole, and deduced that black holes must exist as real objects in the sky around us. They showed that Einstein's theory of general relativity compels any massive star that has exhausted its supply of nuclear fuel to enter a state of permanent free fall. Permanent free fall was a new idea, counterintuitive and profoundly important. It allows a massive star to keep falling permanently into a black hole without ever reaching the bottom.

Einstein never imagined and never accepted this consequence of his theory. Oppenheimer imagined it and accepted it. As a direct result of Oppenheimer's work, we now know that black holes have played and are playing a decisive part in the evolution of the universe. That is the historical fact. The mystery is Oppenheimer's failure to grasp the importance of his own discovery. He lived for twenty-seven years after the discovery, never spoke about it, and never came back to work on it. Several times, I asked him why he did not come back to it. He never answered my question, but always changed the conversation to some other subject.

It is true, as Monk demonstrates, that Oppenheimer's ruling passion was to be a leader in pure science. He considered his excursions into bomb-making and nuclear politics to be temporary interruptions. My interactions with Oppenheimer confirm Monk's picture of him. I worked at the Institute for Advanced Study for almost twenty years while Oppenheimer was the director. He rarely talked about politics and almost never about bombs, but talked incessantly about the latest discoveries and puzzles in pure science.

Twice I had a reason to talk with him about bombs. The first occasion came in 1958, when I asked for a leave of absence from the institute to work on a project in California aimed at building a nuclear bomb–propelled spaceship. I told him how happy I was to be putting his bombs to better use than murdering people. He did not share my enthusiasm. He considered the spaceship project to be an exercise in applied science, unworthy of the attention of an institute professor. The only activity worthy of an institute professor was to think deep thoughts about pure science. He grudgingly gave me a leave of absence for one year, making it clear that if I stayed away for longer than a year I would not be coming back.

The second occasion for me to talk with Oppenheimer about bombs came a few years later, when I was chairman of the Federation of American Scientists, a political organization of scientists concerned with weapons and arms control. The federation was opposing the US deployment of tactical nuclear weapons in exposed positions in Europe and Asia. We considered these deployments to be unacceptably dangerous because nuclear-armed troops involved in local fighting could start a nuclear war that would quickly get out of control. When we examined the history of tactical weapons, we learned that Oppenheimer himself had flown to Paris in 1951 to persuade General Eisenhower, then in command of American forces in Europe, that the United States Army needed tactical nuclear weapons to defend Western Europe against a Soviet invasion. Oppenheimer had been enthusiastically promoting the production and deployment of tactical weapons.

After learning this, I went to see Oppenheimer and asked him directly why he had thought that tactical nuclear weapons were a good idea. This time, he answered my question. He said, "To understand why I advocated tactical weapons, you would have to see the air force war plan that we had then. That was the goddamnedest thing I ever saw. It was a mindless obliteration of cities and populations. Anything, even a major ground war fought with nuclear weapons, was better than that."

I understood then how it happened that Oppenheimer came to grief. He was caught in a battle between the army and the air force. The army wanted small bombs to destroy invading armies. The air force wanted big bombs to destroy whole countries. The army wanted fission bombs and the air force wanted hydrogen bombs. Oppenheimer was on the side of the army. That was why he promoted tactical weapons. That was why he opposed the development of the hydrogen bomb.

The air force took its revenge on the army by helping to drive Oppenheimer out of the government. Air Force General Roscoe Charles Wilson was one of the witnesses against Oppenheimer at the security hearing. General Wilson said, "I felt compelled to go to the Director of Intelligence to express my concern over what I felt was a pattern of action [on Oppenheimer's part] that was simply not helpful to national defense." In the eyes of the air force, anyone who opposed the hydrogen bomb was opposing national defense. The air force won the battle, and Oppenheimer's friends in the army could not help him. The hydrogen bomb development rushed ahead with the highest priority. But in the end, both the air force and the army got all the bombs that they wanted.

Two facts about Oppenheimer stand out clearly from the public record. He was astonishingly effective as the leader of the Los Alamos project. And he never regretted his role as the chief architect of the bomb. In the memoirs of people who worked at Los Alamos, we find many descriptions of his ability to oversee a huge variety of technical jobs, to find the appropriate tools for each pair of hands, and to keep an army of prima donnas working together harmoniously. General Groves talked with many of the leading scientists before choosing one of them to be the director of the project. He chose Oppenheimer because he was the only one with a burning ambition to get the job done. Oppenheimer understood that the project was not scientific but military. Late in 1944, when some of the Los Alamos scientists seemed to be more interested in scientific experiments than in weaponry, Oppenheimer wrote in a memorandum to Groves: "The laboratory is operating under a directive to produce weapons. This directive has been and will be rigorously adhered to."

Oppenheimer continued for the rest of his life to be proud of his achievement at Los Alamos. We know this because he protested vigorously in 1964 when the German playwright Heinar Kipphardt wrote a play portraying him as a tragic hero regretting his actions. Oppenheimer threatened to sue Kipphardt and the producers of the play if they continued to misrepresent him. The producers cut out the offending passages from the play, and the case never went to court.

Oppenheimer continued to block later attempts to produce the play in London and New York. The play was mostly based on the security hearings of 1954. Oppenheimer said in a public statement about the hearings to a *Washington Post* reporter, "The whole damn thing was a farce, and these people are trying to make a tragedy out of it."

Oppenheimer particularly objected to some passages in the play that made him appear anti-American. Monk expresses his opinion, with which I agree, that Oppenheimer's anger arose from his deep loyalty to America. For him, expressing regret for what he had done for his country would have meant joining his country's enemies.

Oppenheimer was above all a good soldier. That is why he worked so well with General Groves, and that is why Groves trusted him. I have a vivid memory of the ice-cold February day in 1967 when we held a memorial service for Oppenheimer at Princeton. Because of the extreme cold, attendance at the service was sparse. But General Groves, old and frail, came all the way from his home to pay his respects to his friend.

I often wondered how it happened that Oppenheimer changed his character so suddenly, from the left-wing bohemian intellectual at Berkeley to the good soldier at Los Alamos. I believe that an important clue to this change is the story of Joe Dallet. In his autobiographical statement at the security hearings, Oppenheimer said:

> It was in the summer of 1939 in Pasadena that I first met my wife.... I learned of her earlier marriage to Joe Dallet, and of his death fighting in Spain.... When I met her I found in her a deep loyalty to her former husband.

After Oppenheimer married Kitty in 1940, they continued to live with the ghost of Dallet. Later I learned from the historian Richard Polenberg some facts about Dallet's life and death.

Dallet was unlike the majority of the left-wing intellectuals who flocked to Spain to fight for the republic. Dallet took soldiering seriously. He believed in discipline. He quickly became an expert on the repair, maintenance, and use of machine guns. He drilled his troops with old-fashioned thoroughness, making sure that they knew how to take care of their weapons and how to use them effectively. In an anarchic situation, his unit was conspicuously well organized. His men caught from him the habit of competence, the pride of a steelworker who knew how to handle machinery. At moments of relaxation, he talked mostly about his beloved machine guns. This was the image of Joe that his friends brought to Kitty in Paris when they came to see her after his death. This was the image that Kitty brought to Oppenheimer when she married him.

From Spain to Los Alamos was a short step. Oppenheimer was as proud of his bombs as Dallet had been proud of his guns. Oppenheimer became the good soldier that Kitty loved and admired. Through the Los Alamos years and for twenty years afterward, the spirit of Dallet lived on in Oppenheimer.

The real tragedy of Oppenheimer's life was not the loss of his security clearance but his failure to be a great scientist. For forty years he put his heart and soul into thinking about deep scientific problems. With the single exception of the collapse of massive stars at the end of their lives, he did not solve any of these problems. Why did he not succeed in scientific research as brilliantly as he succeeded in soldiering and administration? I believe the main reason why he failed was a lack of *Sitzfleisch. Sitzfleisch* is a German word with no equivalent in English. The literal translation is "sitflesh." It means the ability to sit still and work quietly. He could never sit still long enough to do a difficult calculation. His calculations were always done hastily and often full of mistakes. In a letter to my parents quoted by Monk, I described Oppenheimer as I saw him in seminars:

> He is moving around nervously all the time, never stops smoking, and I believe that his impatience is largely beyond his control.

In addition to his restlessness, Oppenheimer had another quality, emphasized by Monk in the subtitle of his book. He always wanted to be at the center. This quality is good for soldiers and politicians but bad for original thinkers. He paid too much attention to famous people working on fashionable topics, while ignoring less famous people working away from the mainstream of science. He had abundant opportunities to learn from two unfashionable geniuses: Fritz Zwicky and John Wheeler. Zwicky was working at the California Institute of Technology throughout the thirteen years when Oppenheimer was a regular visitor. Wheeler was working at Princeton University throughout the twenty years when Oppenheimer was living in Princeton. Zwicky was the discoverer of dark matter, the mysterious invisible stuff that outweighs the visible universe, and he was also a pioneer in the study of supernova explosions and neutron stars. Wheeler was the leading expert on black holes, and the founder of the field of science now known as relativistic astrophysics.

Although Oppenheimer lived close to each of them for many years and knew what they were doing, he did not take their work seriously. He seems to have considered them unworthy of his attention because they were out of the mainstream. Hufbauer reports that Oppenheimer disliked Zwicky and for that reason never used Zwicky's term "neutron star" for the collapsed remnant of a supernova explosion. Wheeler was one of the most enthusiastic proponents of the hydrogen bomb, and Oppenheimer never used Wheeler's term "black hole" for the remnant of a nonexplosive gravitational collapse. Oppenheimer's attitude toward Zwicky and Wheeler was personal antipathy combined with professional misjudgment. As a result, he failed to grasp the opportunities that a closer contact with Zwicky or Wheeler would have provided to make revolutionary discoveries in areas of science ignored by the fashionable mandarins.

Late in Oppenheimer's life, when he was sick and depressed, his wife, Kitty, came to me with a cry for help. She implored me to collaborate with Robert in a piece of technical scientific work. She said Robert was desperate because he was no longer doing science, and he needed a collaborator to get him started. I agreed with Kitty's diagnosis, but I had to tell her that it was too late. I told her that I would like to sit quietly with Robert and hold his hand. His days as a scientist were over. It was too late to cure his anguish with equations.

Note added in 2014: Several readers complained that I wrote a personal story about Oppenheimer and did not review Monk's book. I plead guilty. To me, the parts of Oppenheimer's life and work that Monk includes in his book are less interesting than the parts that he leaves out. I am grateful to Ann Finkbeiner for calling my attention to the book Black Holes and Time Warps *by Kip Thorne (Norton, 1994), which contains in chapter 6 a detailed account of the antagonistic relationship between Oppenheimer and Wheeler. Thorne was present at meetings when Wheeler first vehemently opposed Oppenheimer's theory of black holes and then vehemently supported it, while Oppenheimer made it clear that he did not care what Wheeler thought about it.*

19

HOW TO BE AN UNDERDOG, AND WIN

THE RAND CORPORATION is a think tank in Santa Monica, California, where scholars from many disciplines work for the Department of Defense, mixing academic research with practical advice concerning military problems. The experts at RAND consider themselves the brains of the military establishment. Two fat documents were among those that I read when I visited the RAND Corporation during the years of the American war in Vietnam. One was a magnum opus in six volumes with the title *Oregon Trail*, sponsored by the US Army and written by a large group of historians, many of them eminent university professors. The other was a single volume with the title *Rebellion and Authority*, written by two economists, Nathan Leites and Charles Wolf, working at the RAND Corporation.

The two works were discussing the same problem, made urgent by the situation in Vietnam, of a big and powerful country fighting a weak but determined enemy. The problem was later given the name "asymmetric warfare." Both studies were trying to elucidate the strategy of asymmetric warfare. They reached diametrically opposite conclusions.

The historians writing *Oregon Trail* looked in detail at a hundred examples of asymmetric wars, most of which were colonial wars with a large and wealthy imperial power fighting a group of native rebels. Examples that they examined in depth were the American war of independence, the French colonial wars in Algeria and Vietnam, and the British colonial wars in South Africa and Malaya. Their purpose was to find the general pattern of such wars, to understand why the rebels sometimes won and sometimes lost. They found that the outcome was determined more by psychological than by military factors. Most of the wars lasted between five and ten years, and they usually ended because one side or the other lost the willpower to keep on fighting.

The most important conclusion of the *Oregon Trail* study was that the rebels usually won if the empire spent most of its effort on military operations, but that the rebels usually lost if the empire spent most of its effort on political and social responses to grievances. It was obvious to anyone who read *Oregon Trail* that the American war in Vietnam was likely to be a losing proposition. Unfortunately, very few people had a chance to read it. By one of the worst abuses of the secrecy system that I ever encountered, the military authorities stamped the whole thing secret. By keeping it secret, they made sure that it had no influence on public discussion of the conduct of the war in Vietnam. So far as I know, it remains secret today. Meanwhile, *Rebellion and Authority* was published openly with the blessing of the Department of Defense. It has become a widely accepted guidebook for armies occupying foreign territory and dealing with insurgency.

Forty-five years later, Malcolm Gladwell has written *David and Goliath: Underdogs, Misfits, and the Art of Battling Giants*, another book about asymmetric warfare, beginning with the combat between David and Goliath recorded in chapter 17 of the first book of Samuel.* He describes many examples of asymmetric conflict, in civilian life as

*Little, Brown, 2013.

well as in warfare. He reaches conclusions similar to those of *Oregon Trail*, telling stories that are now fortunately available for all of us to read. Although he had no access to *Oregon Trail*, he has studied *Rebellion and Authority* and explains why he disagrees with it. He quotes a sentence that summarizes the thinking of the economists: "Influencing popular behavior requires neither sympathy nor mysticism, but rather a better understanding of what costs and benefits the individual or the group is concerned with, and how they are calculated." To this piece of economic wisdom, he adds the comment: "In other words, getting insurgents to behave is fundamentally a math problem."

Gladwell goes on to describe the struggle of the British army to subdue the rebellion of the Catholic minority in Northern Ireland, as the economists would see it:

> If there are riots in the streets of Belfast, it's because the costs to rioters of burning houses and smashing windows aren't high enough.... If you were in a position of power, you didn't have to worry about how lawbreakers *felt* about what you were doing. You just had to be tough enough to make them think twice.

Gladwell's main conclusion is that the outcome of an asymmetric conflict depends on legitimacy. The stronger side wins if it can persuade the weaker side that the authority of the stronger side is legitimate. The weaker side wins if it can maintain a firm belief that the stronger side's behavior is illegitimate. This conclusion is expressed in different language by the authors of *Oregon Trail*, but the practical implications of Gladwell and *Oregon Trail* are the same. If the stronger side tries to crush the weaker side with physical force, the stronger side loses legitimacy; the weaker side becomes more determined to resist and usually wins. If the stronger side works hard to redress grievances, it gains legitimacy; the die-hard resisters become isolated and usually lose. The tragedies of Vietnam and Northern Ireland were to some extent a consequence of the fact that the censors gave a voice to the authors of *Rebellion and Authority* and silenced the authors of *Oregon Trail.*

Each of Gladwell's ten chapters carries the name of an underdog. An underdog is a person who struggles with disadvantages in the game of life. Only four of them, including David, were engaged in resistance against superior physical power. The other six were overcoming obstacles in civilian pursuits. Typical of the peaceful heroes is Vivek Ranadivé, who found himself unexpectedly serving as coach of a basketball team of twelve-year-old girls in Redwood City, California. His daughter was a member of the team and persuaded him to volunteer for the job. His disadvantage was the fact that he had grown up in India playing soccer and knew nothing at all about basketball. Because of his ignorance, he trained his girls to play the game like soccer players, constantly running fast after the ball and giving the opposing team no chance to take a breath. This was quite different from the customary way of playing basketball, which has the players concentrating their attention on defending the basket rather than on running. Vivek's team trained hard and played hard, and soon began to beat the other teams who had superior skills but inferior endurance.

Gladwell compares the Redwood City girls with Lawrence of Arabia's team of rebellious Arab tribesmen who beat the Turkish army of occupation in 1917. He quotes Lawrence describing how his tribesmen traveled through the snake-infested desert to attack the Turks in the port city of Aqaba: "Three of our men died of bites; four recovered after great fear and pain, and a swelling of the poisoned limb. Howeitat treatment was to bind up the part with snake-skin plaster, and read chapters of the Koran to the sufferer until he died." Then Gladwell continues:

> When they finally arrived at Aqaba, Lawrence's band of several hundred warriors killed or captured twelve hundred Turks and lost only two men. The Turks simply had not thought that their opponent would be crazy enough to come at them from the desert.

The Redwood City girls beat all the other local teams and ended up playing in the national championships. In the nationals they won their first two games, but then they ran into disaster. The third game was in a town where public opinion was bitterly hostile and the referee was unfair. The referee penalized them repeatedly, declaring their style of play to be illegal, and the public was enthusiastically on the side of the referee. Ranadivé understood that his girls had lost their legitimacy and there was no way they could win. He told them to play the way the referee wanted them to play. As a result, they lost the game and the championship. In peace as in war, the underdog does not always win.

Gladwell emphasizes three inconvenient truths that make the life of underdogs difficult. First, in order to win, underdogs must be disagreeable. The strength of character that enables them to fight against heavy odds makes them insensitive to other people's feelings. Second, they must be prepared to lie and cheat and swindle when necessary. It often happens that they can only escape from bad situations by lying and cheating and swindling. Third, they must be prepared to die for their cause. It frequently happens that they do not live to see their causes prevail.

These three truths are exemplified in several of the stories that Gladwell tells. Ranadivé is the only one of the underdog fighters who is unquestionably a nice guy, and in the end he does not win. Even David, the innocent shepherd-boy hero of the Goliath story, has his dark side. A few years after his victory, he becomes king and steals the wife of his servant Uriah the Hittite. When she gets pregnant, he arranges for Uriah to be killed in battle. When the baby dies, he refuses to mourn. "Unscrupulous" is a good word to describe underdog fighters in general and David in particular. David achieves his purpose in the end when his stolen wife gives birth to Solomon and supplies a legitimate heir for his kingdom.

Wyatt Walker is the hero of a chapter describing the battle between civil rights protesters and segregationist authorities in 1963 in the city of Birmingham, Alabama. This was a classic example of the underdog as trickster, cheating and making mischief in order to win. Walker was second-in-command to Martin Luther King Jr. in the Southern Christian Leadership Conference. Walker organized operations on the ground while King attracted the attention of the world outside. Their strategy was based on two principles. First, provoke the enemy to violent actions that will horrify the world outside and destroy the legitimacy of the authorities. Second, never hit back. Make sure that all protests remain nonviolent and are seen by the outside world to be nonviolent.

Walker had a problem with carrying out the strategy. He had only twenty-two protesters, and it was difficult to provoke the authorities or to attract worldwide attention with such a small number. He played two tricks to make a small number look big. The first trick was to announce a protest march and then delay the start until a large number of spectators came out onto the streets to watch. At that point the television cameras and reporters could not tell the difference between protesters and spectators. The newspapers on the following day reported that eleven hundred protesters had marched. The second trick was to invite all the black high school children in the city to skip school and join the parade. Many hundreds of children came, prepared with freedom signs and singing freedom songs.

After some days of increasing crowds and increasing chaos, the authorities did what Walker had intended them to do. They tried to disperse the crowd by turning high-power fire hoses and police dogs onto the children. A picture appeared on television and in newspapers all over the world, showing a vicious dog attacking a nonviolent black teenager. The teenager was in fact a spectator, not a protester, and he was not hurt. Walker said afterward, "A picture is worth a thousand words." His strategy succeeded, and the result was the passage of the Voting Rights Act two years later, enforcing the right of blacks to vote in elections and ultimately overturning the political power of white segregationists in southern states.

To wage a long campaign of nonviolent resistance, underdog rebels need strict discipline and self-control, and they need a leader with the charisma of King. If the leadership is weak or divided, it is easy for nonviolent resistance to slide into violence and for violence to slide into terrorism. Violence means doing physical harm to wielders of power, such as soldiers or politicians. Terrorism means doing physical harm to innocent bystanders or to whole populations. As a rule, nonviolent tactics give legitimacy to resistance and terrorist tactics give legitimacy to oppressive government. Another inconvenient truth about underdogs is that many of them are terrorists.

Asymmetric wars are usually small wars, fought between a big country and a colony or a group of rebels. But big wars may also be asymmetric. World War I and World War II were big wars, and they were both in important ways asymmetric. World War I was asymmetric if we look at it from the point of view of the man who started it, Gavrilo Princip. Princip was a Bosnian Serb, belonging to a small group of underdogs who were resisting the power of the Austro-Hungarian Empire, which ruled Bosnia. He assassinated Archduke Franz Ferdinand and his wife, Sophie, when they drove through Sarajevo on June 28, 1914. To kill the archduke was an act of resistance. To kill Sophie was an act of terrorism.

Princip started the war and he won it. He achieved both of his grand objectives: the total destruction of the Austro-Hungarian Empire and the independence of the kingdom of Yugoslavia. Yugoslavia united his homelands Bosnia and Serbia in a confederation of Slav peoples. He did not even pay with his life for his victory. He was first imprisoned by the Austrians and then transferred by them to a hospital where he died peacefully of tuberculosis. From the point of view of Princip, the war was a complete success, and the deaths of a few tens of millions were only collateral damage. The war was also a complete success from the point of view of another group of underdog rebels, the Bolsheviks, who took advantage of the war to achieve their aims in Russia.

World War II was asymmetric in a different way. It was started in Europe by Germany and in Asia by Japan, as a conventional war to be fought by big armies on the ground. The aim of Germany was to fight World War I over again and this time win. The aim of Japan was to complete the conquest of China without interference from the United States. The war became asymmetric because Britain and the United States were determined not to fight World War I over again. Britain and the United States made the decision, before the war started, to build large bomber forces that could destroy the enemy homelands from the air.

Germany and Japan did not build strategic bomber forces. The bombing of London was done in a haphazard way by forces not designed for the purpose. The German V-1 and V-2 bombardments were too little and too late to have any substantial effects. Whether the bombing of Germany and Japan was militarily effective is still a matter of dispute. One fact that is not in dispute is that the British and American peoples supported the bombing campaigns, partly for military reasons but mainly to teach the enemy populations a lesson that they would not forget.

Both the Germans and the Japanese had fought all their earlier wars in other people's countries, and now they would finally feel the horrors of war on their own skins. The Germans called the firebombing of their cities *Terrorangriffe*, terror attacks, and they were right. The British public knew that they were terror attacks and was willing to pay the price: 40,000 bomber crewmen dead.

Now, seventy years later, we can see clearly that terrorism worked. In 1945, the year when spectacular firestorms raged in Dresden and Hiroshima, something happened in Germany and Japan that was more profound than military defeat. The traditional cultures of Germany and Japan, which had been the most militaristic on earth, changed abruptly to become the most pacifistic on earth. The change was deep and lasting. Terrorism did not defeat the German and Japanese armies. The Russian and American armies did that. Terrorism did something more difficult and more permanent. It cured the German and Japanese insanities. Terrorism is shock treatment of the crudest sort, but it sometimes works when all else fails.

Gladwell's book is not about big wars and big history. It is about individual people and their problems. In addition to those that I have mentioned, there are seven more underdogs with a chapter for each. They are real people and Gladwell brings them wonderfully to life. The book is divided into three sections. The first is called "The Advantages of Disadvantages (and the Disadvantages of Advantages)." After Ranadivé comes Teresa DeBrito, a schoolteacher who is now principal of the Shepaug Valley Middle School in Connecticut. Her problem is a shortage of kids. The Shepaug Valley has been so gentrified that families with young children can no longer afford to live there. Nearby is the elite Hotchkiss private school, where parents pay exorbitant fees to have their children taught in small classes. DeBrito's classes will soon be smaller than those at Hotchkiss. Parents and politicians think that smaller classes mean better education. But

DeBrito knows from her experience as a working teacher that bigger classes are usually better. One of the best classes she ever taught had twenty-nine kids. The moral of the story is: Things that appear to be disadvantages often turn out to be advantages, and vice versa.

The middle section is called "The Theory of Desirable Difficulty." It begins with David Boies, who is an underdog because he is dyslexic. He struggled through high school and then enjoyed life as a construction worker. Building houses did not require reading. Now he is a famous trial lawyer in California. He says he is a good trial lawyer because he listens. His dyslexia is an advantage because he trained himself to learn everything by listening. He listens to the opposing lawyers and to the witnesses in trials and remembers every word they say. Remembering every word gives him the upper hand.

Emil Freireich had a horrible childhood in extreme poverty in Chicago. During his career as a doctor he was fired seven times for bad behavior. But he devoted his life to finding a cure for childhood leukemia. Leukemia was then a leading cause of death in children. The leukemia ward was a gruesome place soaked in blood, with children in terminal stages bleeding to death. Freireich worked there for twenty years and is largely responsible for the fact that childhood leukemia is now a curable disease. To find the cure and prove that it worked, he had to inflict pain on a lot of children, breaking rules and antagonizing his colleagues. To be tough helped. Freireich said to Gladwell, "I was *never* depressed. I *never* sat with a parent and cried about a child dying.... As a doctor, you swear to give people hope. That's your job."

The last section is called "The Limits of Power," and begins with Rosemary Lawlor, who was a young mother in Belfast when the Troubles began in 1969. The British army imposed a curfew on the Lower Falls area of Belfast, and the people there were running out of food. An army of mothers, pushing prams filled with bread and milk, broke the curfew. Lawlor describes how it happened. "We got the hair pulled out of us. The Brits just grabbed us, threw us up against the walls. Oh, aye! They beat us, like." And then the tide turned. "Once all the people started coming out of their houses, the Brits lost control.... The Brits gave up.... We forced and we forced—until we got in, and we got in and we broke the curfew.... *We did it.*"

The final chapter belongs to André Trocmé, the pastor of the village Le-Chambon-sur-Lignon, which saved the lives of hundreds of Jewish refugees in France under the German occupation. One of the Jews saved was Pierre Sauvage, who was born in the village during the war. He later became a film producer in Hollywood and made a famous documentary film, *Weapons of the Spirit*, with some of the original villagers on screen, describing how the saving of Jews came about. The villagers were ordinary people, living lives of hardship and doing what they thought was right. Gladwell concludes: "It was not the privileged and the fortunate who took in the Jews in France. It was the marginal and the damaged, which should remind us that there are real limits to what evil and misfortune can accomplish." Trocmé was marginal and damaged. He saved the Jews in the village but lost his son. He wrote afterward: "I am like a decapitated pine. Pine trees do not regenerate their tops. They stay twisted, crippled."

Note added in 2014: In the published review I said that Oregon Trail *was a RAND Corporation document. In fact it was an army report, and its official designation is* Project OREGON TRAIL Final Report, USACDC No. USC-6, February 1965. Volume 1, Main Report, TOP SECRET RD. *The full report consisted of two parts: the historical part, which was the biggest part and ought to have been published separately, and a war-game part, which described war games carried out at RAND and at Research Analysis Corporation that were legitimately secret. The decision that I am protesting is the decision to lump the two parts together and classify the whole package as TOP SECRET. As a result of this classification, both parts remain inaccessible after forty-nine years. I am grateful to Lon Jones and to Ashutosh Jogalekar for letters stimulating me to dig out these facts. As usual, I learn from my mistakes only after the review is published.*

20

CHURCHILL: LOVE & THE BOMB

CHURCHILL'S BOMB IS the story of a love triangle.* The three characters are Winston Churchill the statesman, H. G. Wells the writer, and Frederick Lindemann the scientist. Churchill was in love with war and weapons, ever since he was a small boy playing with a historic collection of toy soldiers. Wells wrote books about war and weapons, real and imaginary. Lindemann invented weapons and enjoyed trying them out. War and weapons brought the three of them together. But Churchill could only listen to one guru at a time. The chief source of Churchill's ideas about the application of science to war was Wells in World War I and Lindemann in World War II. Lindemann and Wells, being rivals in love, had nothing but contempt for each other.

Churchill was deeply involved in the prehistory of the atomic bomb for forty years before the bomb existed. More than any other politician, and more than any of the leading scientists of that time, he took seriously the possibility of nuclear weapons. He was born with a romantic attachment to soldiering, enjoyed applying high technology

*Graham Farmelo, *Churchill's Bomb: How the United States Overtook Britain in the First Nuclear Arms Race* (Basic Books, 2013).

to military problems, and found kindling for his imagination in the science-fiction stories of Wells.

His personal friendship with Wells began in 1901, when he read Wells's nonfiction work *Anticipations* and responded with an eight-page fan letter. The friendship lasted until Wells's death in 1946. Churchill reacted enthusiastically to Wells's book *The War in the Air*, which appeared in 1908 with vivid descriptions of the military uses of the newly invented airplane. In January 1914 Wells published *The World Set Free*, a story that gave starring roles to two new inventions: "land ironclads," later known as tanks, and "atomic bombs," later known as nuclear weapons. Churchill pushed the development and use of tanks in World War I. He understood that they would give soldiers a chance to break out of the horrors of the trenches, making warfare quick and mobile. His tanks came too late to get the boys out of the trenches in that war, but they arrived in time to have a decisive effect in World War II. He gave full credit to Wells for the idea.

Churchill's thinking about nuclear weapons was summarized in a piece, "Fifty Years Hence," published in *Strand Magazine* in 1931. "There is no question among scientists," he wrote,

> that this gigantic source of energy exists. What is lacking is the match to set the bonfire alight....The busy hands of the scientists are already fumbling with the keys of all the chambers hitherto forbidden to mankind....Without an equal growth of mercy, pity, peace and love, science herself may destroy all that makes human life majestic and tolerable.

The match to light the nuclear fire was the fission of uranium, discovered in 1938 by Otto Hahn and Fritz Strassmann in Berlin.

Lindemann worked at the Royal Aircraft Factory at Farnborough during World War I and became famous for solving the problem of tailspin. Many pilots were losing their lives because their aircraft would stall during combat maneuvers, fall into a tailspin, and helplessly spin into the ground. Lindemann worked out the theory of tailspin and found a remedy. He calculated that the pilot could give a counterintuitive push to the rudder, which would convert the spin into a straight dive and allow the pilot to regain control. He then borrowed an airplane, put it into a tailspin, applied the push that he had calculated, pulled out of the straight dive, and flew the plane safely home. This combination of scientific wizardry and courage won him the lifelong admiration of Churchill.

Lindemann met Churchill for the first time in 1921 and explained recent scientific discoveries in simple language. Churchill found him to be a kindred spirit, an old-fashioned patriot who saw no shame in using science to win wars. In 1924, Churchill wrote an essay about the future of warfare with the title "Shall We All Commit Suicide?," describing apocalyptic visions of anthrax weapons and "a bomb no bigger than an orange... [with] a secret power to concentrate the force of a thousand tons of cordite and blast a township at a stroke." Before writing the piece, he turned for advice to Lindemann and not to Wells.

Still Wells remained faithful to his old love. In 1908 he had written a piece for the *Daily News*, "Why Socialists Should Vote for Mr. Churchill." In 1940 he wrote a piece for *Collier's* magazine, "Churchill, Man of Destiny." His verdict on Churchill in 1940: "He has pulled himself together. He is pulling us all together. It is like awakening from a nightmare to think of what might have happened to my country without him." When the chips were down, Wells was an old-fashioned patriot too.

Wells was a spinner of fanciful tales while Lindemann was a real scientist. Paradoxically, the information that Wells gave to Churchill was mostly right, while the information that Lindemann gave was mostly wrong. Wells had been right about airplanes and tanks before World War I. Lindemann was wrong about radar in 1935, when it was first proposed for defending Britain against attack from the air. He gave low priority to radar, which turned out to be the decisive technology of World War II and was crucial to the defense of Britain in 1940. One of the offshoots of radar was the proximity fuse, which enabled an antiaircraft shell to destroy an aircraft without hitting it directly. The proximity fuse multiplied the kill rate of antiaircraft artillery by a factor of ten. In 1944, when the V-1 drone airplanes were attacking London, a massive line of antiaircraft guns with proximity fuses was deployed along the coast and succeeded in shooting down 70 percent of the V-1s before they reached England. If the Germans had had proximity fuses for their antiaircraft guns, they could probably have stopped our large-scale bombing of Germany.

Lindemann gave the highest priority to aerial mines. Aerial mines were his pride and joy. The idea was to destroy airplanes with mines floating in the air, just as ships were destroyed by mines floating in the water. The big difference between air and sea is that the air has three dimensions while the surface of the sea has two dimensions. An aerial mine has to kill airplanes over a wide range of heights. The mine with the explosive charge must hang at the bottom of a long steel wire with a parachute at the top. If an airplane flies into the wire, the wire will bite into the skin of the wing until it reaches solid metal. Pulled upward by the drag of the parachute, the wire will slide up through the wing until the explosive charge reaches the airplane and detonates. Lindemann continued to play with this toy all through the years of World War II. It absorbed a large amount of money and attention that might have been put to better use.

It was obvious to almost everyone except Lindemann that aerial mines could not be an effective defense. The wire had to be thousands of feet long and correspondingly heavy. Even with a big parachute, it would not stay in the air for more than a few minutes. To defend an important target, a fleet of airplanes would be required to continue sowing mines over the area as long as the attack continued. If many targets were to be defended, the defense would quickly run out of mines. And it was easy to invent countermeasures. A system of small clippers along the leading edge of an airplane wing could cut the wires and make aerial mines harmless.

When I was working for the British Bomber Command toward the end of World War II, we would from time to time receive inquiries from some high level of government, asking whether damage to returning bombers gave any evidence that the Germans were using aerial mines. Our answer was always negative. My boss told me confidentially that the inquiries were coming from Lindemann.

Lindemann was enthusiastic about technical toys such as aerial mines, but he remained unenthusiastic about nuclear weapons. One week after the beginning of World War II, he moved from Oxford to London to become a full-time scientific adviser to Churchill, who was then First Lord of the Admiralty. Lindemann was well aware of the discovery of fission and the possibility of nuclear weapons, but he waited for two years before advising Churchill to begin a project to develop a British bomb. Toward the end of the war, Lindemann visited the American bomb laboratory at Los Alamos and remarked privately to his friend Reginald Jones, "What fools the Americans will look after spending so much money." Jones had been Lindemann's student before the war, and worked closely with him as the head of scientific intelligence. Jones said that until the bomb exploded at Alamogordo, Lindemann never really believed that the thing would work.

The title *Churchill's Bomb* is misleading. It was probably chosen by the publisher to attract readers rather than to describe the book. Graham Farmelo's main subject is the personal rivalry surrounding the British nuclear weapons project, in which Churchill played a leading part. But the book is not a history of the bomb. It does not answer some of the obvious questions that a reader might ask: What were the technical obstacles to be overcome? What did the scientists actually do while the politicians argued about it? How was the bomb built? How was it supposed to be delivered? What effect has it had? Is the bomb still relevant in the world of today, sixty years after it was built? Why is it called Churchill's bomb rather than Attlee's bomb? After all, it was Clement Attlee and not Churchill who gave the order to build it.

The subtitle, "How the United States Overtook Britain in the First Nuclear Arms Race," is also misleading. There was never an arms race between the United States and Britain. There was an arms race between Britain and Germany, beginning in 1939 and ending in 1942. During that time the United States was still neutral and not seriously engaged in the race. Britain won the race when Werner Heisenberg and Albert Speer secretly agreed to abandon the German nuclear bomb project. Then, in 1942, with the United States at war, Britain and the United States still believed that they were in a race with Germany, since they did not know that the Germans had given up. The choice for Britain was whether to join forces with the United States or to try to build a bomb independently.

Churchill made the decision to merge British efforts with the American project. A merger meant sharing secrets, and the sharing of secrets was always a delicate problem. A year went by before sharing became effective and British scientists were working at Los Alamos. During that year, the American project took a great leap forward and the British project stalled. Enrico Fermi with his American colleagues built the first nuclear reactor in Chicago and explored the new world of nuclear power. British scientists spent the year waiting for the American authorities to allow them to participate. It was true that the United States overtook Britain, but Churchill was not racing. Churchill had already decided that he wanted a partnership with America and not a race.

The nuclear partnership began in 1943 and came to a sudden end with the passing of the McMahon Act by Congress in 1946. That year, the United States had bombs and the industrial equipment to make more bombs, and Britain was shut out. Britain had to decide whether to give up or go ahead with building an independent British bomb. Attlee had taken Churchill's place as prime minister in 1945 and made the decision to go ahead with the British bomb. It was successfully tested in 1952, when Churchill was back in power. In that same year the Americans tested the first hydrogen bomb with a yield of ten megatons. Churchill quietly gave the order for a British hydrogen bomb, which was built and successfully tested in 1957. By that time Churchill had ended his second term as prime minister, but he achieved his goal of restoring the nuclear partnership and the sharing of secrets with America.

Churchill, Lindemann, and Wells did not fundamentally disagree about nuclear strategy. They agreed that nuclear weapons were desirable as instruments of power, immensely dangerous, and historically decisive. Churchill and Lindemann saw the bomb as necessary to preserve the status of Britain as a great power. Wells saw it as necessary to establish the authority of a future world government.

Only one voice spoke out in well-reasoned opposition to these views. The opposing voice belonged to Patrick Blackett, a physicist who had served as a naval officer in World War I, survived the Battle of Jutland in 1916, and led the team of scientists helping the Royal Navy to defeat German U-boats in World War II. He won a Nobel Prize in 1948 for discoveries in particle physics. Both as a scientist and as an expert in war-fighting, Blackett had far better credentials than Lindemann. But Blackett was a socialist and was active in left-wing politics. Lindemann hated him, and Churchill distrusted him.

They made sure that Blackett was kept out of all high-level discussions of nuclear policy so long as Churchill was prime minister.

As soon as Attlee became prime minister in 1945, he appointed Blackett to his Advisory Committee on Atomic Energy. The next year was the decisive turning point in the history of nuclear weapons. Several governments made serious proposals to put the newly created United Nations in charge of the nascent nuclear industries all over the world, with power to prevent any nation from building nuclear bombs. This was the last chance to avoid a large-scale nuclear arms race. Robert Oppenheimer in the United States and Niels Bohr in Denmark were the leaders of a worldwide campaign of scientists for international control of nuclear energy. The United Nations Atomic Energy Commission was created to exercise whatever form of international control the member nations could agree to establish. Everything depended on finding an international legal frame that the United States and the Soviet Union could both accept.

The US proposal for international control was known as the Baruch plan because it was written by Bernard Baruch, a conservative banker and a friend of Churchill. The essential point that made it unacceptable to the Soviet Union was the enforcement clause, which gave the United Nations Security Council power to enforce the agreement by majority vote. In all other actions of the Security Council, each permanent member of the council had the right to veto majority decisions. In the Baruch plan the right to veto was abolished for decisions concerning nuclear weapons. In any dispute involving the Soviet Union, the Soviet Union was likely to be in the minority and the United States in the majority, so the Baruch plan was giving a permanent nuclear hegemony to the United States. Oppenheimer fought hard inside the American government for a plan that would recognize the Soviet need for equal treatment. Blackett fought hard inside the British government. Oppenheimer failed to convince Truman and

Blackett failed to convince Attlee. American hegemony was what both Truman and Attlee wanted and hoped to make permanent.

Stalin knew that the American hegemony would not last long. He said, "The atomic bomb is a good weapon for threatening people with weak nerves." Stalin did not have weak nerves. He knew that his country had produced more tanks than Germany in wartime and could produce more atomic bombs than the United States in peacetime. In 1946 the Soviet Union proposed a simple prohibition of nuclear weapons, overseen by the United Nations but without any enforcement clause. After a year of argument about details, the negotiations ended and the nuclear arms race began. The American hegemony ended with the first Soviet bomb test in 1949.

Blackett disagreed strongly with Attlee, not only about the Baruch plan but also about the decision to build a British bomb. Blackett believed that the military value of the bomb was illusory while the danger of possessing it was real. He argued that the bomb would be useless in any future wars that Britain might reasonably fight. Any war that was worth fighting could be won with nonnuclear weapons. And if there were ever a nuclear war involving the Soviet Union, the possession of nuclear weapons would make it sure that London and other British cities would be obliterated.

After Blackett failed to find support for these views inside the government, he made them public in a book that was published in Britain with the title *Military and Political Consequences of Atomic Energy* and in America with the title *Fear, War, and the Bomb*. The book appeared in 1948 and became a best seller with translations published in eleven languages. Farmelo says rightly that the book had no influence on government policies or on majority opinions at the time. He says wrongly that the book is "so dense that much of it is barely readable." In fact it is highly readable and widely read. It stands after sixty-five years as a classic statement of the case against the nuclear follies of our age. Some of Blackett's predictions have been proved wrong and some of his arguments have become irrelevant, but the central theme of his book is still true. He is saying that the military utility of the bomb is small, that its political importance is exaggerated, and that only its danger as an instrument of mass murder is real.

Blackett said in 1948 that the Soviet proposal for abolishing nuclear weapons without enforcement should have been accepted. If his advice had been followed, we would have been in a situation like the one in 1972 when the United States, the United Kingdom, and the Soviet Union signed a treaty abolishing biological weapons. Before the treaty was signed, the large stockpiles of American and British biological weapons had been destroyed. After the treaty was signed, the Soviet Union cheated on a massive scale and continued to maintain a large clandestine stockpile. Today the biological weapons treaty is still in force and we still have reason to suspect that Russia may be cheating. The question now is whether we are better off with the treaty or without it. Is it better to have a world with biological weapons illegal and well hidden in clandestine facilities, or a world with large stockpiles of biological weapons openly deployed and vulnerable to theft?

Opinions may be divided on the value of the treaty, but there is at least a reasonable argument to be made for keeping it in force. The same argument was made by Blackett for accepting the 1946 Soviet proposal for prohibiting nuclear weapons. If we had accepted the Soviet proposal, we would be living in a world with nuclear weapons legally prohibited but secretly manufactured and hidden away in various places around the world. Would that world be less dangerous than the world of huge stockpiles openly deployed in which we have lived for the last sixty years? Blackett answered yes to that question.

It is time now for the world to ask the question again and decide whether Blackett was right.

Looking back with seventy years of hindsight, we can see clearly that Churchill was deluded. Central to his vision of the world was the power and glory of the British Empire. He fought his wars for the preservation of the empire. The young people who fought for Britain in World War II were not fighting for the empire. They knew that the empire was crumbling and most of them were happy to see it swept away. That was why they voted in 1945 to sweep Churchill away. They knew that Churchill was living in the past, out of touch with the real world. I have a vivid memory of the British general election of 1950, when Attlee was running for reelection after five years of slow recovery from the war. Attlee came to Birmingham, where I was then living, to give a campaign speech to a large crowd. He spoke at length about the social programs that the Labour Party had carried out during his tenure, the big improvements in public housing and public education, and the National Health Service. The crowd listened to this without much enthusiasm. Then at the end of his speech, Attlee said, "We gave freedom to India," and the crowd responded with loud and long cheering. Giving freedom to India was the one thing that Churchill would never have done. Attlee won the election.

When Churchill returned for his second term as prime minister, he recognized that the empire was fading and based his nuclear policy on another illusion, the special relationship between Britain and America. During World War II he had enjoyed a special relationship with Franklin Roosevelt, with frequent telephone calls and many personal meetings. His friendship with Roosevelt was a crucial part of his war strategy. It allowed him to think of himself as one of the Big Three, deciding the fate of the world in conferences with Roosevelt and Stalin. After returning to office in 1951, he tried to reestablish his special relationship with presidents Truman and Eisenhower. Truman and Eisenhower found his personal advances annoying and gently pushed him off. After the British hydrogen bomb was demonstrated in 1957, sharing of nuclear secrets was successfully reestablished, but Churchill's belief that this would perpetuate Britain's status as a Great Power remained an illusion.

The big question that Farmelo does not try to answer is whether it makes sense for Britain to have nuclear weapons. Two famous scientists answered this question with a resounding no. One was Patrick Blackett. The other was Joseph Rotblat, a Polish nuclear physicist who went with the British contingent to Los Alamos. Rotblat was the only scientist who left the bomb project in 1944 as soon as he heard that the Germans were not working on a bomb. He served for most of a long life as leader of the Pugwash movement, an international alliance of scientists concerned about war and weapons. For his efforts as a peacemaker he won the Nobel Peace Prize in 1995.

Just as Rotblat is unique as a Los Alamos scientist who walked out of the brotherhood of bomb-makers for reasons of conscience, the Republic of South Africa is unique as a country possessing nuclear weapons and unilaterally destroying them. The South Africans have set a splendid example for other countries possessing nuclear weapons to follow. Nobody gives South Africans diminished respect because they walked out of the nuclear club. The United Kingdom is now in an excellent position to gain respect and save money by following the South African example.

21

THE CASE FOR BLUNDERS

SCIENCE CONSISTS OF facts and theories. Facts and theories are born in different ways and are judged by different standards. Facts are supposed to be true or false. They are discovered by observers or experimenters. A scientist who claims to have discovered a fact that turns out to be wrong is judged harshly. One wrong fact is enough to ruin a career.

Theories have an entirely different status. They are free creations of the human mind, intended to describe our understanding of nature. Since our understanding is incomplete, theories are provisional. Theories are tools of understanding, and a tool does not need to be precisely true in order to be useful. Theories are supposed to be more or less true, with plenty of room for disagreement. A scientist who invents a theory that turns out to be wrong is judged leniently. Mistakes are tolerated, so long as the culprit is willing to correct them when nature proves them wrong.

Brilliant Blunders, by Mario Livio,* is a lively account of five wrong theories proposed by five great scientists during the last two

**Brilliant Blunders: From Darwin to Einstein—Colossal Mistakes by Great Scientists That Changed Our Understanding of Life and the Universe* (Simon and Schuster, 2013).

centuries. These examples give for nonexpert readers a good picture of the way science works. The inventor of a brilliant idea cannot tell whether it is right or wrong. Livio quotes the psychologist Daniel Kahneman describing how theories are born: "We can't live in a state of perpetual doubt, so we make up the best story possible and we live as if the story were true." A theory that began as a wild guess ends as a firm belief. Humans need beliefs in order to live, and great scientists are no exception. Great scientists produce right theories and wrong theories, and believe in them with equal conviction.

The essential point of Livio's book is to show the passionate pursuit of wrong theories as a part of the normal development of science. Science is not concerned only with things that we understand. The most exciting and creative parts of science are concerned with things that we are still struggling to understand. Wrong theories are not an impediment to the progress of science. They are a central part of the struggle.

The five chief characters in Livio's drama are Charles Darwin, William Thomson (Lord Kelvin), Linus Pauling, Fred Hoyle, and Albert Einstein. Each of them made major contributions to the understanding of nature, and each believed firmly in a theory that turned out to be wrong. Darwin explained the evolution of life with his theory of natural selection of inherited variations, but believed in a theory of blending inheritance that made the propagation of new variations impossible. Kelvin discovered basic laws of energy and heat, and then used these laws to calculate an estimate of the age of the earth that was too short by a factor of fifty. Pauling discovered the chemical structure of protein, the active component of all living tissues, and proposed a completely wrong structure for DNA, the passive component that carries hereditary information from parent to offspring.

Hoyle discovered the process by which the heavier elements essential for life, such as carbon, nitrogen, oxygen, and iron, are created by nuclear reactions in the cores of massive stars. He then proposed a theory of the history of the universe known as steady-state cosmology, which has the universe existing forever without any big bang at the beginning, and stubbornly maintained his belief in the steady state long after observations proved that the big bang really happened.

Finally, Einstein discovered the great theory of space and time and gravitation known as general relativity, and then added to the theory an additional component later known as dark energy. Einstein afterward withdrew his proposal of dark energy, believing that it was unnecessary. Long after Einstein's death, observations have proved that dark energy really exists, so that Einstein's addition to the theory was right and his withdrawal was wrong.

Each of these examples shows in a different way how wrong ideas can be helpful or unhelpful to the search for truth. No matter whether wrong ideas are helpful or unhelpful, they are in any case unavoidable. Science is a risky enterprise, like other human enterprises such as business and politics and warfare and marriage. The more brilliant the enterprise, the greater the risks. Every scientific revolution requires a shift from one way of thinking to another. The pioneer who leads the shift has an imperfect grasp of the new way of thinking and cannot foresee its consequences. Wrong ideas and false trails are part of the landscape to be explored.

Darwin's wrong idea was the blending theory of inheritance, which supposed the qualities inherited by offspring to be a blend of the qualities of the parents. This was the theory of inheritance generally accepted by plant breeders and animal breeders in Darwin's time. Darwin accepted it as a working hypothesis because it was the only theory available. He accepted it reluctantly because he knew that it was unsatisfactory in two ways. First, it failed to explain the frequent cases of hereditary throwback, when a striking hereditary feature such as red hair or musical talent skips a generation from grandparent to grandchild. Second, it failed to allow a rare advantageous variation to spread from a single individual to an entire population of animals, as required by his theory of the origin of species. With blending inheritance, any rare advantageous variation would be quickly diluted in later generations and would lose its selective advantage. For both of these reasons, Darwin knew that the theory of blending inheritance was inadequate, but he did not have any acceptable alternative when he published *The Origin of Species* in 1859.

Nine years later, when Darwin published another book, *The Variation of Animals and Plants Under Domestication*, he had abandoned the blending inheritance theory as inconsistent with the facts. He replaced it with another theory that he called pangenesis. Pangenesis said that the inheritance of qualities from parent to offspring was not carried in the seeds alone but in all the cells of the parent. Somehow the cells of the parent produced little granules that were collected by the seeds. The granules then instructed the seeds how to grow. For the rest of his life Darwin continued to believe in pangenesis, but it was another brilliant blunder, no better than blending inheritance and equally inconsistent with the facts.

Like Darwin's theories of blending heredity and pangenesis, Kelvin's wrong calculation of the age of the earth and Pauling's wrong structure for DNA were speculations requiring blindness to obvious facts. Kelvin based his calculation on his belief that the mantle of the earth was solid and could transfer heat from the interior to the surface only by conduction. We now know that the mantle is partially fluid and transfers most of the heat by the far more efficient process of convection, which carries heat by a massive circulation of hot rock moving upward and cooler rock moving downward. Kelvin lacked our modern knowledge of the structure and dynamics of the earth, but he could see with his own eyes the eruptions of volcanoes bringing hot liquid from deep underground to the surface. His skill as a calculator seems to have blinded him to messy processes such as volcanic eruptions that could not be calculated.

Similarly, Pauling guessed a wrong structure for DNA because he assumed that a pattern that worked for protein would also work for DNA. He was blind to the gross chemical differences between protein and DNA. Francis Crick and James Watson, paying attention to the differences, found the correct structure for DNA one year after Pauling missed it.

Hoyle's wrong theory of the universe had a different status from the other mistakes, because he was a young rebel when he proposed it. The steady-state universe was from the beginning a minority view. The decisive evidence against it was the discovery in 1964 of the microwave radiation pervading the universe. The microwave radiation had been predicted to exist as a relic of the hot big bang. The radiation proved that the hot big bang really happened and that the universe had a violent beginning. After that discovery, Hoyle was almost alone, preaching the steady-state gospel to a small band of disciples.

Einstein, the last of Livio's five blunderers, is an exception to all rules. He is widely quoted as saying that his addition of dark energy to the theory of gravitation was his biggest blunder. Livio has carefully examined the evidence and has come to the conclusion that Einstein never made this statement. The evidence points strongly to George Gamow as the guilty party. Gamow was another brilliant blunderer with a reputation for making up colorful stories. Einstein's real biggest blunder happened when he changed his mind and dropped dark energy from his theory. Nature proved him wrong fifty years after his death, when it revealed that three quarters of the total mass of the universe is dark energy.

Einstein invented a steady-state model of the universe many years before Hoyle. This steady-state model was discovered recently by a group of Irish scientists in an unpublished Einstein manuscript. Einstein abandoned the idea and never published it, probably because he found that steady-state theories are contrived and artificial. When Hoyle noisily promoted the steady-state cosmology twenty years later, Einstein never mentioned that he had discovered and discarded it long before. Einstein must have recognized it quickly as a brilliant blunder, clever but not likely to be correct. (I am indebted to the Irish scientist Cormac O'Raifeartaigh for information about the discovery of the Einstein manuscript.)

After reading Livio's account, I look on the history of science in a new way. In every century and every science, I see brilliant blunders. Isaac Newton's biggest blunder was his corpuscular theory of light, which had light consisting of a spray of little particles traveling along straight lines. In the nineteenth century, James Clerk Maxwell discovered the laws of electromagnetism and proposed that light consists of electromagnetic waves. In the twentieth, Einstein proved that Newton and Maxwell were both right and both wrong, because light behaves like particles in one situation and like waves in another.

The chief difference between science and other human enterprises such as warfare and politics is that brilliant blunders in science are less costly. Hannibal's brilliant crossing of the Alps to invade Italy from the north resulted in the ruin and total destruction of his homeland. Two thousand years later, the brilliant attack on Pearl Harbor cost the Japanese emperor his empire. Even the worst scientific blunders do not do so much damage.

The worst political blunder in the history of civilization was probably the decision of the emperor of China in the year 1433 to stop exploring the oceans and to destroy the ships capable of exploration and the written records of their voyages. In no way can this blunder be called brilliant. Before the decision, China had a fleet of ocean-going ships bigger and more capable than any European ships. China was roughly level with Europe in scientific knowledge and far ahead in the technologies of printing, navigation, and rocketry. As a consequence of the decision, China fell disastrously behind in science and technology, and is only catching up now after six hundred years. The decision was the result of powerful people pursuing partisan squabbles and neglecting the long-range interests of the empire. This is a disease to which governments of all kinds, including democracies, are dangerously susceptible.

Another cause of catastrophic blunders is religion. A legendary example of a religious blunder is the story of Tsar Lazar, the king of Serbia in 1389 when his kingdom was invaded by the Turks. He confronted the Turkish army on the fatal battlefield of Kosovo Polje. The story is told in the Serbian national epic *The Battle of Kosovo*. The Virgin Mary happened to be in Jerusalem at the time when the Turks invaded, and sent a falcon with a message for the tsar. The falcon arrived on the battlefield and told the tsar that he must make a choice between an earthly and a heavenly kingdom. If he chose the earthly kingdom, his army would defeat the Turks and he would continue his reign in Serbia. If he chose the heavenly kingdom, his army would be annihilated and his people would become slaves of the Ottoman Empire. Being a very pious monarch with his mind concentrated on spiritual virtue, the tsar naturally chose the heavenly kingdom, and his people paid for his choice by losing their freedom.

Seven years after Darwin published *The Origin of Species*, without any satisfactory explanation of hereditary variations, the Austrian monk Gregor Mendel published his paper "Experiments in Plant Hybridization" in the journal of the Brünn Natural History Society. Mendel had solved Darwin's problem. He proposed that inheritance is carried by discrete units, later known as genes, that do not blend but are carried unchanged from generation to generation.

The Mendelian theory of inheritance fits perfectly with Darwin's theory of natural selection. Mendel had read Darwin's book, but Darwin never read Mendel's paper.

The essential insight of Mendel was to see that sexual reproduction is a system for introducing randomness into inheritance. In garden peas as in humans, each plant is either male or female, and each offspring has one male and one female parent. Inherited characteristics may be specified by one gene or by several genes. Single-gene characteristics are the simplest to calculate, and Mendel chose them to study. For example, he studied the inheritance of pod color, determined by a single gene that has a version specifying green and a version specifying yellow. Each plant has two copies of the gene, one from each parent. There are three kinds of plants, pure green with two green versions of the gene, pure yellow with two yellow versions, and mixed with one green and one yellow. It happens that only one green gene is required to make a pod green, so that the mixed plants look the same as the pure green plants. Mendel describes this state of affairs by saying that green is dominant and yellow is recessive.

Mendel did his classic experiment by observing three generations of plants. The first generation was pure green and pure yellow. He crossed them, pure green with pure yellow, so that the second generation was all mixed. He then crossed the second generation with itself, so that the third generation had all mixed parents. Each third-generation plant had one gene from each parent, with an equal chance that each gene would be green or yellow. On the average, the third generation would be one-quarter pure green, one-quarter pure yellow, and one-half mixed. In outward appearance the third generation would be three-quarters green and one-quarter yellow.

This ratio of 3 between green and yellow in the third generation was the new prediction of Mendel's theory. Most of his experiments were designed to test this prediction. But Mendel understood very well that the ratio 3 would only hold on the average. Since the offspring chose one gene from each parent and every choice was random, the numbers of green and yellow in the third generation were subject to large statistical fluctuations. To test the theory in a meaningful way, it was essential to understand the statistical fluctuations. Fortunately, Mendel understood statistics.

Mendel understood that to test the ratio 3 with high accuracy he would need huge numbers of plants. It would take about eight thousand plants in the third generation to be reasonably sure that the observed ratio would be between 2.9 and 3.1. He actually used 8,023 plants in the third generation and obtained the ratio 3.01. He also tested other characteristics besides color, and used altogether 17,290 third-generation plants. His experiments required immense patience, continuing for eight years with meticulous attention to detail. Every plant was carefully isolated to prevent any intruding bee from causing an unintended fertilization. A monastery garden was an ideal location for such experiments.

In 1866, the year Mendel's paper was published, but without any knowledge of Mendel, Darwin did exactly the same experiment. Darwin used snapdragons instead of peas, and tested the inheritance of flower shape instead of pod color. Like Mendel, he bred three generations of plants and observed the ratio of normal-shaped to star-shaped flowers in the third generation. Unlike Mendel, he had no understanding of statistical fluctuations. He used a total of only 125 third-generation plants and obtained a value of 2.4 for the crucial ratio. This value is within the expected statistical uncertainty, either for a true value of 2 or for a true value of 3, with such a small sample of plants. Darwin did not understand that he would need a much larger sample to obtain a meaningful result.

Mendel's sample was sixty-four times larger than Darwin's, so that Mendel's statistical uncertainty was eight times smaller. Darwin failed to repeat his experiment with a larger number of plants, and missed his chance to incorporate Mendel's laws of heredity into his theory of evolution. He had no inkling that a fundamental discovery was within his grasp if he continued the experiment with larger populations. The basic idea of Mendel was that the laws of inheritance would become simple when inheritance was considered as a random process. This idea never occurred to Darwin. That was why Darwin learned nothing from his snapdragon experiment. It remained a brilliant blunder.

Mendel made a brilliant blunder of a different kind. He published his laws of heredity, with a full account of the experiments on which the laws were based, in 1866, seven years after Darwin had published *The Origin of Species*. Mendel was familiar with Darwin's ideas and was well aware that his own discoveries would give powerful support to Darwin's theory of natural selection as the cause of evolution. Mendelian inheritance by random variation would provide the raw material for Darwinian selection to work on.

Mendel had to make a fateful choice. If he chose to call Darwin's attention to his work, Darwin would have understood its importance, and Mendel would inevitably have become involved in the acrimonious public disputes that were raging all over Europe about Darwin's ideas. If Mendel chose to remain silent, he could continue to pursue his true vocation, to serve his God as a monk and later as the abbot of his monastery. Like Tsar Lazar five hundred years earlier, he had to choose between worldly fame and divine service. Being the man he was, he chose divine service. Unfortunately, his God played a cruel joke on him, giving him divine gifts as a scientist and mediocre talents as an abbot. He abandoned the chance to be a world-famous scientist and became an unsuccessful religious administrator.

Darwin's blindness and Mendel's reticence combined to delay the progress of science by thirty years. But thirty years is a short time in the history of science. In the end, after both men were dead and their personal shortcomings forgotten, their partial visions of the truth came together to create the modern theory of evolution. Thomas Hunt Morgan at Columbia University understood that the fruit fly *Drosophila* was a far better tool than the garden pea and the snapdragon for studying heredity. Fruit flies breed much faster and are more easily handled in large numbers. With fruit flies, Morgan could go far beyond Mendel in exploring the world of genetics.

In my own life as a scientist, there was one occasion when I felt that a deep secret of nature had been revealed to me. This was my personal brilliant blunder. I remember it with joy, even though my dreams of glory were shattered. It was a blissful experience. It arose out of work that I did with my colleague Andrew Lenard from Indiana University, investigating the stability of ordinary matter. We proved by a laborious mathematical calculation that ordinary matter is stable. The physical basis of stability is the exclusion principle, a law of nature saying that two electrons can never be in the same state. Matter is stable against collapse because every atom contains electrons and the electrons resist being squeezed together.

My blunder began when I tried to extend the stability argument to other kinds of particles besides electrons. We can divide particles into two types in three different ways. A particle may be electrically charged or neutral. It may be weakly or strongly interacting. And it may belong to one of two types that we call fermions and bosons in honor of the Italian physicist Enrico Fermi and the Indian physicist Satyendra Bose. Fermions obey the exclusion principle and bosons do not. So each particle has eight possible ways to make the three choices. For example, the electron is a charged weak fermion. The light quantum is a neutral weak boson. The famous particle predicted by Peter Higgs, and discovered in 2012 at the European Center for Nuclear Research (CERN), is a neutral strong boson.

I observed in 1967 that seven of the eight possible combinations were seen in nature. The one combination that had never been seen was a charged weak boson. The missing type of particle would be like an electron without the exclusion principle. Next, I observed that our proof of the stability of matter would fail if electrons without the exclusion principle existed. So I jumped to the conclusion that a charged weak boson could not exist in a stable universe. This was a new law of nature that I had discovered. I published it quietly in a mathematical journal.

I knew that my theory flatly contradicted the prevailing wisdom. The prevailing wisdom was the unified theory of weak and electromagnetic interactions proposed by my friends Steven Weinberg and Abdus Salam. Weinberg and Salam predicted the existence of a new particle as a carrier of weak interactions. They called the new particle W. The W particle had to be a charged weak boson, precisely the combination that I had declared impossible. Nature, speaking through an experiment at CERN in Geneva, would decide who was right.

The decision did not come quickly. It took the experimenters fifteen years to build a new machine and use it to search for the W particle. But the decision, when it came, was final. Large numbers of W particles were seen, with the properties predicted by Weinberg and Salam. With hindsight I could see several reasons why my stability argument would not apply to W particles. They are too massive and too short-lived to be a constituent of anything that resembles ordinary matter. I quickly forgot my disappointment and shared the joy of Weinberg and Salam in their well-deserved triumph. As my mother taught me long ago, the key to enjoyment of any sport is to be a good loser.

In Livio's list of brilliant blunderers, Darwin and Einstein were good losers, Kelvin and Pauling were not so good, and Hoyle was the worst. The greatest scientists are the best losers. That is one of the reasons why we love the game. As Einstein said, God is sophisticated but not malicious. Nature never loses, and she plays fair.

Note added in 2014: Several readers wrote indignant letters complaining because I accused great scientists of blundering. I replied to them as follows:

> Thank you for your comments on my review. I learn more from people who disagree with me than from those who agree.... The main cause of disagreement in this case is a different understanding of the word "blunder." To me and to Mario Livio the word carries no blame. We use it in a joking fashion to refer to any proposal or opinion that turns out to be wrong or ignorant. To you the word seems to be judgmental, as if committing a blunder were committing a crime. To me, the main value of Mario Livio's book is to make blunders respectable.

SOURCES

The essays in this book were originally published as follows:

"Our Biotech Future," *The New York Review of Books*, July 19, 2007

The letter responding to "Our Biotech Future" was published in *The New York Review of Books*, September 27, 2007, and is reprinted here with the permission of Wendall Berry.

"Writing Nature's Greatest Book," *The New York Review of Books*, October 19, 2006

"Rocket Man," *The New York Review of Books*, January 17, 2008

The letter responding to "Rocket Man" was published in *The New York Review of Books*, February 14, 2008, and is reprinted here with the permission of Bernard Lytton.

"The Dream of Scientific Brotherhood," *The New York Review of Books*, May 10, 2007

"Working for the Revolution," *The New York Review of Books*, October 25, 2007

"The Question of Global Warming," *The New York Review of Books*, June 12, 2008

The letter responding to "The Question of Global Warming" was published in *The New York Review of Books*, October 9, 2008, and is reprinted here with the permission of Robert M. May.

"Struggle for the Islands," *The New York Review of Books*, October 23, 2008

"Leaping into the Grand Unknown," *The New York Review of Books*, April 9, 2009

The letter responding to "Leaping into the Grand Unknown" is reprinted here with the permission of Frank Wilczek.

"When Science and Poetry Were Friends," *The New York Review of Books*, August 13, 2009

"What Price Glory?," *The New York Review of Books*, June 10, 2010

The letter responding to "What Price Glory?" is reprinted here with the permission of Steven Weinberg.

"Silent Quantum Genius," *The New York Review of Books*, February 25, 2010

"The Case for Far-Out Possibilities," *The New York Review of Books*, November 10, 2011

"Science on the Rampage," *The New York Review of Books*, April 5, 2012

"How We Know," *The New York Review of Books*, March 10, 2011

"The 'Dramatic Picture' of Richard Feynman," *The New York Review of Books*, July 14, 2011

"How to Dispel Your Illusions," *The New York Review of Books*, December 22, 2011

The letter responding to "How to Dispel Your Illusions" was published in *The New York Review of Books*, January 12, 2012, and is reprinted here with the permission of Daniel Kahneman.

"What Can You Really Know?," *The New York Review of Books*, November 8, 2012

"Oppenheimer: The Shape of Genius," *The New York Review of Books*, August 15, 2013

"How to Be an Underdog, and Win," *The New York Review of Books*, November 21, 2013

"Churchill & the Bomb," *The New York Review of Books*, April 24, 2014

"The Case for Blunders," *The New York Review of Books*, March 6, 2014